Did you enjoy this issue of BioCoder?

Sign up and we'll deliver future issues and news about the community for FREE.

http://oreilly.com/go/biocoder-news

BioCoder

WINTER 2014

O'REILLY® Beijing · Cambridge · Farnham · Köln · Sebastopol · Tokyo

Contents

BioCoder and Education

Nina DiPrimio, PhD

Thanks to those of you who are returning to BioCoder, and welcome to the new readers and contributors. The first issue of BioCoder reached more than 10,000 readers, and we hope that number continues to grow. We've put together an exciting second issue and think you'll find the diverse content to be educational and inspirational.

I taught laboratory courses for a couple years after earning my PhD, but I then decided to dive back into research. However, I still wanted to teach, but I wanted to teach individuals who chose the topics they wanted to learn and who attended the classes for their own personal development. That led to the creation of IdeaLab at BioCurious in Sunnyvale, California: a meeting to discuss current research topics, analyze scientific articles, and generate project ideas. After a successful year of the class, we joined forces with Genspace and created a virtual journal club open to all individuals with the desire to learn about current topics in research. Within the independent scientist community, this is one example of many ways in which individuals are learning research fundamentals, broadening their areas of knowledge and expertise, and sparking new project ideas and collaborative efforts.

What does this have to do with BioCoder?

BioCoder is another forum to educate those who are willing to learn, but with this venue, we have the opportunity to reach a larger audience. This means that we can inspire those who aren't involved in biology to get involved, make science less intimidating and more accessible, enhance collaboration, and maybe even affect science policy. I challenge you to think about ways in which you can continue to spread the knowledge that will affect public opinion. We are doing amazing things in biology, bioengineering, and related areas of research, and you'll see many examples of that in this issue. The content ranges from scientific education through museum exhibitions and bioart to an exploration of entrepreneurial topics for the independent scientists, including tips on crowdfunding your project, biosafety rules and regulations, biological gaming, and open source healthcare.

We will continue to include diverse content in BioCoder issues, but in order to fulfill that promise, we need to hear from you, our readers. If you have a unique perspective on scientific happenings outside of traditional settings—or maybe untraditional scientific happenings in traditional settings—please contact us at *biocoder@oreilly.com*. We would love to hear your ideas for future articles.

Enjoy, explore and continue to educate!

DIYbio and the Hacking Metaphor

Michael Scroggins

Within DIYbio, one cannot escape the hacking metaphor. The metaphor is ubiquitous and, to a point, useful. The term connotes both productive play with an existing technology aimed at improvement and, at the same time, play with sinister undertones. In this sense, hacking captures the promise and pitfalls of the dual uses any mature technology might be put to, whether that technology is as dramatic as nuclear power/weapons or as mundane as a free/premium software license. But every metaphor has its limits. Pushed too far, metaphors break down, and instead of illuminating, they obscure. Which brings me to ask: how far can the hacking metaphor be pushed within DIYbio—at least the part of DIYbio falling in line with synthetic biology?

Nowhere has the hacking metaphor been taken as literally as Andrew Hessel, Marc Goodman, and Steven Kotler took it in the 2012 *Atlantic* article, "Hacking the President's DNA." Hessel, Goodman, and Kotler imagine a world in which biology is a mature technology and maps one-to-one with computer programming. A world where novices can create human viruses and infect individuals with them in just the same way the unskilled can infect computers with malware. They craft a dramatic narrative around this metaphor by imagining the president of the United States as the target of sinister biohackers. The authors justify the premise of their article by arguing that the technology to make the scenario plausible is accelerating at an exponential rate and that it will only be a few short years before it is mature and ubiquitous. Therefore, better to prepare now than suffer tomorrow.

But is this the case? Is biology a few years away from the scenario outlined in *The Atlantic*? Two years ago, I attended a symposium on the current state and future of synthetic biology. It was the usual mix of technical talks and hallway conversations, but a question asked in the plenary roundtable stood out. During the Q&A, a young undergraduate asked a synthetic biologist the following question:

Undergrad: *At Google, an entry-level job requires a BA, but in synthetic biology, an entry-level job requires a postdoc. Should this be rethought?*

Biologist: *Have you looked into DIYbio?*

The undergrad brushed off the biologist's response as beside the point, and the testy exchange ended shortly thereafter with the biologist pointing out that there was much basic research to be done in creating tools for the commons. But the undergrad's question remains hanging. What kind of training counts here? There is an established discourse holding that synthetic biology—and, by extension, much of DIYbio—is analogous to computer science, but the implication of the student's question and the biologist's deflection argue for a different answer. The question posed to the biologist has two parts, both of which shed light on the scenario envisioned by Hessel, Goodman, and Kotler: a) why must so much synthetic biology be done *in vivo* or *in vitro*, not *in silico*, thus making necessary a long and arduous course of laboratory training? and b) why doesn't the work translate directly to the *in-vivo* world outside the laboratory and thus bring about an industry comparable to the software industry?

In answering an earlier question posed during the roundtable, the biologist picked up on the tool comment when he said there was basic work in metrology to be done in synthetic biology before the academic work could be readily translated into the commercial realm. This has been true in almost all theoretical fields that have been translated into commercial ventures. The fact that synthetic biology is still coming to terms with metrology means that it has many years of difficult laboratory work ahead and that the scenario envisioned by Hessel, Goodman, and Kotler is more fear mongering than an accurate assessment of the near-future possibilities.

Michael Scroggins is a PhD candidate at Teachers College, Columbia, and a researcher at the Center for Everyday Education. He is currently in Silicon Valley conducting research for the project "Education into Technological Frontiers: Hackerspaces as Educational Institutions." He can be reached via email at michaeljscroggins@gmail.com or at http://about.me/michael_scroggins.

From Closed Exhibits to Open Labs

The Continued Evolution of Science Museums

Oliver Medvedik

Growing up in New York City, I have vivid memories of taking class trips during grade school to the American Museum of Natural History. Located right across the street from Manhattan's equally historic Central Park, I remember the breathtakingly realistic dioramas of stuffed and mounted wildlife, representative of every continent. Posed within meticulously crafted environments that captured their habitat and behavior in a three-dimensional snapshot, they were encased behind walls of glass. Despite their continued haunting beauty, they also unwittingly captured forever a past outlook on science's interaction with the public. The exhibits said almost as much about the societal mindset that had created them as they did about the flora and fauna held within. With the work of science hermetically sealed off, only to be gazed at with passive awe, they placed the uninitiated public squarely to one side of the scientific divide.

On few occasions, as young students, we sometimes managed to get a glimpse of the scientific work that went on behind the scenes. Once, we even got to go on a special trip to the geology laboratories that were closed off to the general public where the staff scientists analyzed specimens for the museum. At that time, during the early 1980s, that was as close to interactivity as it got for me at a science museum. However, things have been rapidly changing, and it seems that the spirit of the late '60s has finally penetrated science museums around the globe.

Interactivity at the Exploratorium

But just what is the mission of a museum these days, especially one dedicated to the dissemination of science? According to the Merriam-Webster online dictionary, one definition is "a building in which interesting and valuable things (such as paintings and sculptures or scientific or historical objects) are collected and shown to the public." Simple and to the point as far as storing and displaying artifacts is concerned, but what about when it comes to science and technology museums?

With the advent of the Exploratorium in San Francisco in the summer of 1969, the science museum was effectively reinvented. Started by the physicist Dr. Frank Oppenheimer, the Exploratorium was to be an entirely new way of educating the public. A space where interactivity was the rule, with static hands-off displays being replaced by hands-on exhibits that encouraged the public to tinker, engaging the additional senses of touch and sound.

This radical departure from the traditional, staid museum caught on rapidly, with science, technology, and even math museums around the nation vying to out-do one another in the depths of interactivity. In this era of high-bandwidth con-nectivity, 3D printers, and cheap, fast electronics, the tools at the disposal of the science museum curator offer unprecedented flexibility to design immersive sci-entific exhibits.

Exhibits are one thing, but museums—ever on the lookout for the next wave in the promotion of science—are now opening up their spaces to the teaching of science, technology, and medicine. As one example, the Liberty Science Center, opened in 1993 in New Jersey, has multiple laboratories where students can do projects and take courses. There are even teleconferencing links for students to talk to surgeons live during open-heart surgery!

The American Museum of Natural History also offers a diverse variety of hands-on, after-school research programs available for students, in subjects as di-verse as biotechnology and astronomy. Other museums, such as the New York Hall of Science, along with the Exploratorium on the West Coast, play host to the annual Maker Faire, the wildly popular gathering of makers, doers, tinkerers, and inven-tors.

With all this happening at science museums, is there anything else left for them to do? Have we possibly reached the limits of public involvement in science and technology?

Open Labs Reach Science Museums

The emphasis on citizen science-based initiatives has definitely not escaped notice in science museums around the world. This past month, I was privileged to visit and help organize and run several biotechnology workshops held at the Zhejiang Science and Technology Museum (ZJMST) just south of Shanghai, in the city of Hangzhou in China (see Figure 3-1 and Figure 3-2). Part of its stated mission is to lead the development of science museums in Zhejiang Province and to disseminate scientific knowledge to the public. Along with my friend, neuroscientist Dr. Pia-Kelsey O'Neill from Columbia University, I embarked on the trip at the request of Bing Zhu, the deputy director of the International Department of ZAST, the Zhejiang Association for Science and Technology. ZAST is tasked with helping to promote scientific outreach within the province of Zhejiang. Each province in China has its own Association for Science and Technology, with similar missions. Upon arrival, we were quickly introduced to Mr. Li Ruihong, the director of the science museum. Ever since the initiative to promote interactive exhibits was launched at the Exploratorium, the continued evolution of science museums and the expanding roles that they can play as advocates of science has certainly taken root around the world. After many discussions with staff at ZJMST, it quickly became apparent that they were ready to start moving beyond what had been achieved thus far. Our brainstorming discussions started to envision how open labs could function in spaces such as science museums.

A New Community of Science

Where do I see the evolution of science museums heading? With the flexibility inherent in their mission as public institutions dedicated to the promotion of science and technology to all citizens, I believe that they are uniquely poised to play a central role in helping to launch citizen-science initiatives. They can act as hubs for the planning, gathering, and analysis of data gathered from a variety of crowd-sourced projects. They can serve as a fertile meeting ground between educated amateurs and trained scientists. And not least, like community biotechnology laboratories and hackerspaces, with no inherent publish-or-perish imperative, science can be more freely pursued here in the most positive sense of the term "amateur": for the sheer love of gathering knowledge.

Those who have participated in shared coworking spaces, either community biotech labs or other hacker spaces, have experienced firsthand the need for more spaces that facilitate such interactions. With the addition of community laboratories, a huge step will be taken, not just toward the global accessibility of scientific information, but for the practice of science as well. Will the establishment of open community laboratories within science museums harbor the emergence of a sort of new secular cathedral? Will there be a global race among museums and nations to outdo one another in making the skills, tools, and philosophy of science and technology accessible to everyone as a result? If yes, then this is a trend I will certainly be supporting.

Figure 3-1. Staff of Zhejiang Science and Technology Museum practice pipetting for a multiplex PCR and gel electrophoresis workshop used to identify several species of animals for the purposes of food identification

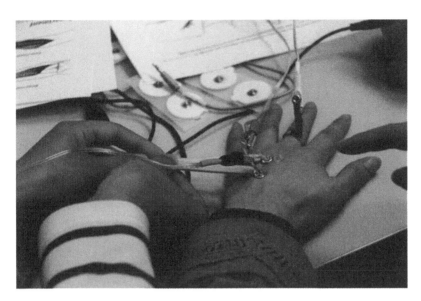

Figure 3-2. Staff of Zhejiang Science and Technology Museum use electrodes placed on the surface of their skin and connected to a SpikerBox (http://www.backyardbrains.com), an inexpensive open source device used to amplify and measure the rates of action potentials traveling through neurons

Oliver Medvedik is currently the Sandholm Visiting Assistant Professor of Biology and Bioengineering at The Cooper Union School of Engineering. He is also a cofounder of Genspace and serves as its scientific director. He resides in New York City.

On Art and Science: The Enlightened Renaissance

Simona Zompi

The studies of art and science were not mutually exclusive until the late 19th century, when specialization became fashionable and separated the two. Science education became even more fashionable in the education system, as it was easier to measure as compared to arts and humanities—it's easy to evaluate if a student has learned math, but harder to evaluate whether she is a good writer or a promising painter.

It was not until the late 20th century, with the boom of design into technology, that the arts regained a space in the innovation field. The explosion of hacker spaces and DIYbio laboratories became fertile ground to intersect art and science, and they're both the consequence and the cradles of what I like to call the "Enlightened Renaissance."

When we think about iconic figures mixing art and science, the first person who comes to mind is Leonardo da Vinci (1452–1519), the archetypal Renaissance man. An artist by training, his curiosity in scientific observation and his accurate records in journals made him a scientist. He made exceptional contributions to the study of human anatomy, being one of the first to draw the internal organs and the development of fetuses in the womb. But more importantly, he linked the observation of the human heart with its physiology, describing how heart valves control the flow of blood, bringing together art and physiology. This was not science based on experimentation or testing of theories, but rather it was a different approach to science: intense observation and detailed recording.

Leonardo initially turned to science in order to improve his artwork; he studied the science of light and the proportions of the human body exemplified in the Vitruvian Man. But who are those scientists who turn to art in order to improve their scientific work?

Intense observation as developed by visual arts can greatly improve scientists' observational skills. Dr. Karen Sokal-Gutierrez, a physician and researcher at UC Berkeley, told me that after she started painting human bodies and skin tones—mixing different pigments (Prussian blue, burnt sienna, cadmium red and yellow, and different greens)—she started examining her patients differently. Exercising her eyes through painting helped her medical practice by developing the range of colors and tones she could see.

Charles Darwin (1809–1882), the most famous natural scientist of his century, drew inspiration from the tradition of natural history and illustration, namely geological and botanical drawings. This closeness to nature is still inspiring today's environmental scientists. In turn, Darwin's theory inspired many artists, especially the French impressionist painters, such as Cezanne and Monet, and influenced the meaning of beauty. Sir Alexander Fleming (1881–1995), the discoverer of the popular antibiotic penicillin, was a scientist and also a painter and a member of the Chelsea Arts Club, a private club for artists. He mostly used watercolors, but he also used live organisms and was the first to experiment with microbial art in his "germ paintings"—Fleming would inoculate petri dishes with different bacterial species to create live paintings that would fade away once the bacteria died (*http://www.microbialart.com/galleries/fleming/*). In order to develop his microbial art, Fleming had to find microbes with different pigments and then time his inoculations such that the different species all matured at the same time. The inspiration for this art would drive the discovery of new microbes.

As he developed intense observation through his art practice, Fleming's discovery of penicillin may not be as fortuitous as some may think. His discovery came about when he accidentally left on his bench several petri dishes plated with the bacteria staphylococcus. After a few weeks, returning from vacation, he noticed that a fungus, penicillium, had contaminated one of the petri dishes and that the bacteria directly surrounding the fungus were dead. Many before him probably had petri dishes contaminated by mold, but his observation skills made him see through the petri dishes and deduce that the mold had killed the bacterial colonies. Among all his germ paintings, he had made a masterpiece that would save millions of lives.

After Fleming, many experimented with living bacteria and art. *Exploring the Invisible* is a collaboration between the artist Ann Brodie and the microbiologist Dr.

Simon Park. They used a strain of bioluminescent bacteria, Photobacterium Phosphoreum, as the source of light in a dark chamber called the bioluminescent photo booth. Volunteers were photographed or recorded in the booth exploring new interactions between microbes and humans (*http://bioproject.tumblr.com/*). The Bioglyphs Project at Montana State Unversity uses bioluminescent bacteria to create all forms of art (*http://www.biofilm.montana.edu/Bioglyphs/*). Dr. Peta Clancy's *Visible Human Bodies* uses inoculated petri dishes to "paint" human bodies (*http://petaclancy.com/works/*) just like Sir Fleming did before her.

John Maeda, president of the Rhode Island School of Design, said that "innovation doesn't just come from equations or new kinds of chemicals, it comes from a human place. Innovation in the sciences is always linked in some way, either directly or indirectly, to a human experience. And human experiences happen through engaging with the arts—listening to music, say, or seeing a piece of art." Scientists are humans after all. He insists on the need to add an "A for art" to the mix of STEM (science, technology, engineering, and math) programs, creating STEAM.

Some schools in California have integrated STEM and arts. At UC Davis, the Art and Science Fusion Program (*http://artsciencefusion.ucdavis.edu/home page.html*) follows Darwin's path, combining natural history, environment, and art. At Stanford, the Senior Reflection (TSR) program (*http://www.stanford.edu/~su emcc/TSR/*), led by Andrew Todhunter and Dr. Susan McConnell, allows senior undergraduates majoring in science to create an art piece based on their favorite science topic. This art piece is meant to be the capstone creation of the students' four years of college. The Center for Science Education at UC Berkeley Space Sciences Laboratory, directed by Dr. Laura Peticolas, integrates art and science by bringing artists in residence into the lab (*http://cse.ssl.berkeley.edu/arts/news.html*). Most recently, three students at UC Berkeley—Natalie Mal, Kimiya Hojjat, and Angela Weinberg—inspired by Stanford TSR, started leading a DeCal course on art and science to explore how the two fields interact and cross-fertilize. The launch of the DeCal earlier this year was a huge success. Many students had to be turned down, and only 16 students were allowed to join the class to keep the group small and interactive. This past December, an art show showcased students' original work. Finally, the Leonardo Art and Science Evening Rendezvous (LASERs) started by Piero Scaruffi in 2008 has now piqued national interest, bringing artists and scientists together to foster interdisciplinary work (*http://www.scaruffi.com/leonardo/*). These meetings are open to the general public and take place in colleges and universities, bringing the public closer to academia.

The Renaissance was a time of wondrous discoveries, when any enquiring mind would be encouraged to pursue scientific discoveries. The Age of Enlightenment, beginning in the late 17th and 18th centuries, advanced knowledge using scientific experimental methods. During this time, scientific publications, salons, and debating societies bloomed all over Europe. Art allows more intuition and a larger perspective; it allows reaching the limits of what is possible, breaking of scientific rules, and creation of new and unique ideas. The hacker's revolution is a return to the Renaissance age mixed with the Age of Enlightenment. It allows anyone to use scientific experimental models mixed with a broader artistic intuition. With hackers' spaces and DIYbio spaces we are entering the age of the "Enlightened Renaissance."

Simona Zompi is a physician and scientist and is currently the director of research at the Center for Youth Wellness in San Francisco, California; board member of Counter Culture Labs in Oakland, California; and an independent consultant in immunology and global health. She is mentoring the students who are leading the DeCal course on art and science at UC Berkeley. As part of her doctoral degree, she studied the molecular pathways activated in murine natural killer cells upon infection and tumor development. Most recently, she worked as a project scientist and led the development of the human immunology component of a dengue program at UC Berkeley in collaboration with the National Virology Lab in Managua, Nicaragua. She has worked in various countries on HIV programs and served as the executive director of the Center for Global Public Health at UC Berkeley.

DIYbio Around the World, Part II

Noah Most

Until August 2014, I will be traveling the globe visiting people within the DIYbio community and playing around with biology. In October, I finished up what was a grand introduction to DIYbio in Victoria, Canada, at the Victoria Makerspace. Among many amazing experiences, I worked on a project to explore the feasibility of DNA origami for DIYbio.

DIY DNA Origami: Wrapping Up from Last Issue[1]

Paul Rothemund dropped a gauntlet in his 2008 TED talk (*http://www.ted.com/talks/paul_rothemund_details_dna_folding.html*), describing his new method in the field of DNA nanotechnology as "so easy you could do it at home in your kitchen and design the stuff on a laptop." That method is called DNA origami, and it enables you to treat DNA as a nanoscale construction material. For example, researchers have used this technique to create nanorobots for the delivery of drugs[2] as well as programmable nano-breadboards.[3] A long, single-stranded DNA "scaffold," typically the M13mp18 bacteriophage, is folded upon itself and pinched into the desired shape by oligonucleotide "staples" that are easily designed on the computer.[4] To

---

1. Most, Noah. "DIYbio Around the World," *BioCoder*, fall 2013.

2. Douglas, Shawn M., Ido Bachelet, and George M. Church. "A Logic-gated Nanorobot for Targeted Transport of Molecular Payloads," *Science* 335.6070, 2012, 831–834.

3. Maune, Hareem T. et al. "Self-assembly of carbon nanotubes into two-dimensional geometries using DNA origami templates," *Nature Nanotechnology*, 5(1), 2009, 61–66.

4. Rothemund, Paul W. K. "Folding DNA to create nanoscale shapes and patterns," *Nature*, 440(7082), 2006, 297–302.

our knowledge, no one has taken Rothemund up on the kitchen biology challenge and actually attempted DIY DNA origami. Last issue, I examined whether this could be a reasonable DIYbio project, and in this issue, I can happily report an answer: yes! We successfully synthesized a proof-of-concept design, a simple 2D light bulb, and imaged it with an atomic force microscope (AFM). In tandem, we produced a crash course in DNA origami that made it so conceptually accessible that even a 12-year-old understood it.

However, that "yes" should be qualified by a discussion of some of the difficulties we encountered while imaging our nanostructure. Although the design and synthesis of DNA origami structures is relatively straightforward, imaging takes some patience and adds further cost. Microscope time will typically cost you per hour. AFM probes cost around $30 each and typically come in packs of no fewer than 10, which is more than should be needed for a single design. (For a review of other major costs, see my article in the fall 2013 issue of BioCoder.[5]) The tip of a new probe can be destroyed in an instant if it is mishandled or dropped. More perplexingly, it took several sittings for us to produce decent AFM images without procedural changes clearly accounting for the difference in image quality. However, some of these problems may stem from using AFM probes with fewer floppy tips than is typical for the method. Finally, documentation of imaging protocol in the literature is often too brief to be useful to a newcomer. As a result, we somewhat blindly tried to figure out everything on our own, before finally, in a magical moment, we saw our tiny light bulb, less than 1,000th the width of a human hair.

These obstacles should not be prohibitive, and most can be avoided. To avoid them, we recommend that you contact us or convince a nearby professional DNA origami lab to let you observe its imaging protocol. While DNA origami remains more expensive than the typical DIY activity, the field oozes excitement because it is almost the Wild West—fresh, unexplored territory. Only a fraction of its potential applications have even been imagined, so give it a try!

Science Gallery, Dublin, Ireland

After Victoria, my original intention to visit the Manchester MadLab abruptly veered off course when presented with an awesome detour to help Dr. Ellen Jorgensen, of Genspace in New York, with an exhibition at the Science Gallery (*https://dublin.scien cegallery.com/*) in Dublin, Ireland. The Science Gallery houses four to six rotating art and science themes per year, and, from October 2013 until January 2014, it's

5. Most, Noah. "DIYbio Around the World," *BioCoder*, fall 2013.

hosting *Grow Your Own*, an exhibition that examines biotechnology like synthetic biology and its implications. Since no discussion of the implications of synthetic biology is complete without mention of DIYbio, Ellen set up a small community lab in the gallery so that gallery-goers could see, and do, synthetic biology themselves.

I hadn't met Ellen in person before, so she welcomed me by interrogation. Hurling rapid-fire questions in my direction, she tested my ability to address the most persnickety and combative journalism. After I weathered her barrage (it was actually fun), Ellen warmed up, I donned a ridiculous tie-dye lab coat, and we set about entertaining and educating visitors about our glass box. Gallery-goers toured the community lab, asked questions about DIYbio and synthetic biology, had their DNA extracted, examined one of their genes by PCR, tried out some wet work (a Genomikon kit) themselves, and/or saw what were, to some, downright terrifying genetically engineered organisms.

One advantage of DIYbio is that it's hands on, which amplifies its power as a communication tool. Genetic engineering is much less intrinsically terrifying when you see it firsthand. An older woman walked into our lab and said, "You don't do any of that genetic engineering stuff, do you?" I grabbed a nearby plate with E. coli that had been engineered to express green fluorescent protein (GFP) from a jelly-fish, and said, stretching it out, "Yes, in fact, we do." Literally jumping backward, she shrieked, "That's dangerous!" Half an hour later, she had gotten much more comfortable with the nonpathogenic, highly attenuated strain on the plate, and maybe even the idea that genetic engineering products are not *necessarily* harmful, irrespective of what was changed. Having Nobel Prize–winning examples, like GFP, in hand certainly helps dispel disproportionate fears.

What are artists saying about the brave new biotech world? The other pieces in the Science Gallery likely raise more fears than they dispel, but in doing so also highlight important questions. For example, Heather Dewey-Hagborg (*https://dublin.sciencegallery.com/growyourown/strangervisions*), who collaborated with Genspace, found items discarded by strangers on the streets and created a portrait of them based on just their DNA, which told her about ancestry, freckles, eye color, etc. The result is remarkably creepy. Anonymous faces stare blankly out from a wall —the source of all the information, a simple cigarette butt, below. However, the implications of the work are even creepier. What does it say about privacy when someone can sense your key physical attributes and whether you have an increased risk of Parkinson's from almost anything you leave behind?

The New Weathermen (*https://dublin.sciencegallery.com/growyourown/newwea thermen*) by David Beneque imagines that a renegade activist group uses biotechnology to advance its environmental aims. They target the African oil palm, which produces the cheap palm oil often found in fast food. Many groups, such as Greenpeace (*http://bit.ly/1aJMC9u*), criticize the expansion of the tree's cultivation because it may be done at the expense of rainforests. The installation diagrams their attack plan. They infect the world's palm oil trees, genetically engineering their oil so that it gives people indigestion. Palm oil trees are now worthless, the rainforests are saved, and the renegades reap an additional reward: "#Biolulz."

Perhaps the piece I enjoyed the most was one that appeared on my plate. The Center for Genomic Gastronomy (*http://genomicgastronomy.com/*) hosted a meal where each of the six courses was designed to make you look not only at your plate, but also at the issues surrounding it, in a new way. For example, I devoured a dish made of invasive organisms, a pigeon standing in for a de-extincted carrier pigeon, and, my personal favorite, ribs with Cobalt 60 BBQ sauce, which largely uses ingredients generated with radiation mutagenesis. This virtually unknown process to rapidly generate mutants with useful traits underpins many of the crops we eat today—some 3,000 have been released—and stands in sharp contrast to genetically engineered foods, which have attracted much greater public scrutiny (*http://bit.ly/ 1cWfvff*). I adored this dinner because it amplified attendees' interest in biotechnology, inspiring them to ask new questions while also giving me an opportunity to share some of what I've learned. Overall, I left the Science Gallery with a newfound appreciation for bioart. Although artists may take certain creative liberties that irk scientists, they perform an essential function, especially for getting people who might not otherwise care interested.

From Dublin, I visited Cathal Garvey, Ireland's famed biohacker, in Cork and received a lovely tour of his basement lab. Since my trip to Ireland, I most recently met the Berlin DIYbio group. Unfortunately, this article can only cover a small fraction of my experiences since the last issue; I want future articles to conglomerate experience by category rather than in a fragmented, quasi-chronological fashion. So, as I hop around and see more, I'll report on European DIYbio next time.

Noah Most graduated last May from Grinnell College, where he studied biology, economics, and entrepreneurship. He has performed research in labs across the country and led a microfinance nonprofit that was hailed as a "Champion of Change" by President Obama. Through an interest in synthetic biology, he discovered DIYbio, and, six months later, he won the Thomas J. Watson Fellowship in order to study it around the world for a year.

Superseding Institutions in Science and Medicine

Anthony Di Franco

Recently, on a mailing list about open source medical devices, Damon Muma asked the following question: "I am feeling the urge to contribute to this movement, but I'm a bit lost on what has been done/tried/planned or what would be effective and useful. Worried that efforts would be duplicating others or not usefully focused. I've had random friends who aren't even diabetic but into coding offer to help but wasn't sure what to point them at."

My response began with some fairly concrete observations based on my experiences as a type I diabetic, but it grew to encompass more fundamental issues I know about as a person with an academic background in systems/control theory who works with machine learning techniques as a programmer. I wanted to share those observations here as a way to provoke thought on these questions and broaden the discussion.

Security in Medical Devices

In 2011, the tragically late Barnaby Jack invited me to share the stage with him a couple times to help him present his insulin pump hacks, where he demonstrated the near-total lack of security in insulin pumps, permitting remote control of all functions with no prior knowledge about an individual pump. He moved on to similar work with pacemakers. I was very much looking forward to his work putting pressure on manufacturers to secure their products, and more broadly to improve their quality standards and innovate a bit in their products' function to produce meaningful progress in patient standards of care (none of which they seem to have much interest in now, certainly not considering the enormous profits they reap from selling the products).

Closed-loop Control in Insulin Pumps

A few months ago, Medtronic began marketing an incrementally improved model of its pump as an *artifical pancreas*. Until now, in every usage I'm aware of, "artificial pancreas" has meant automatic dosing based mainly on continuous sensor readings. This new pump has the same manual dosing as the old models but adds a feature that cuts off insulin if the patient is trending into hypoglycemia. This is no doubt a useful feature for many diabetics, and I am happy to see it come to market, but suggesting that it has much to do with closed-loop blood glucose control is mere hype, and an insult to the diabetic community's intelligence, and, I worry, symptomatic of the broad lack of innovation in megacorporatized medicine.

I suspect this even more because, from my academic background in control theory, I find the control algorithms that appear in the public research I've seen on closed-loop blood glucose control to be some of the oldest and least powerful and stable in the field of control. The proportional-integral-derivative (PID) method still appears prominently in the research, even though in the broader field of control, much more powerful and stable approaches such as model-predictive control have been in routine use for decades, and began to appear in closed-loop blood glucose control research during the last 15 years. PID control is distinguished mainly by having origins in the 1890s and being one of the only techniques possible before the invention of practical computers because it fit into the limitations of pneumatic-powered mechanical governor technology. It appears in the early chapters of many textbooks and remains in widespread use in devices such as thermostats—used to solve simple, one-dimensional problems—but it serves mainly as an example of an inadequate approach when there are any subtleties in the problem to overcome. Closed-loop blood glucose control involves many of the most notable of these subtleties, including nonlinearity, long time lags, and very noisy and biased feedback, and would likely benefit from being formulated in a multiple-input, multiple-state, multiple-output way in order to capture the relevant information and reflect the relevant complexities, instead of the one-dimensional formulation that PID is suited to. (For example, a multiple-input formulation would be necessary to use amylin, insulin, and glucagon infusions together to more fully mimic pancreatic function and prevent blood glucose swings at the beginning and end of the insulin dose response.)

It seems, from the outside, like a poorly supported effort organized more around pursuing minimally viable increments to existing products than around taking the technical problem and the goal of producing excellent outcomes for patients seriously on their own terms. There must be many reasons for this, and I speculate that alongside the petty bureaucratic ones, there are even ones based in concerns I would agree are objectively valid. But it is a problem in itself that this all happens out of sight of the main stakeholders in the results: the diabetic patients —that all that reaches the patients is preposterous hype. And in a context of institutionalized science and research and development, it is hard to imagine any significant change in this situation. Opacity and inertia are the default, and in many senses the main objectives, of this form of organization. Only token concessions can be made against them within the paradigm.

The community of diabetics could work around this apparent bottleneck and aggregate sensor and dose data from patients into an open dataset with which to build models and do offline experiments with proposed algorithms, which is an established methodology in the field. It could then build an ecosystem of open algorithms and hold contests for their improvement and selection. This would become a valuable resource for both institutional and citizen research efforts, and an important resource for checking scientific validity by reproducing results in a field hindered by datasets being mostly proprietary.

Open Source Treatments

Open sourcing everything required to treat diabetes is the most ambitious and difficult goal. However, it synergizes well with the broader DIYbio movement for two reasons. First, because recombinant insulin was the first major commercial success of biotechnology and set the pattern for future development of biologics. Second, because the tools involved are fundamental to biology and medicine and overlap well with the toolset needed for all serious DIYbio research and other community goals: production and purification of biologics, infusion pumps, and in-vivo sensing.

Open sourcing treatments is also important because pharmaceuticals in general and insulin pumping specifically suffer from very perverse economic incentives that, at least in part, favor keeping people on the treatments with the most disposable or consumable supplies possible, at the highest price relative to the cost of manufacturing. Open source efforts could develop incentives more attuned to encouraging innovations that improve patient outcomes than to encouraging

Man, insulin withdrawal is *horrible!*

Figure 6-1. The effects of untreated diabetes include blindness, impotence, nerve damage and necrosis in the extremities, kidney failure, cardiovascular disease, coma, and death—all good reasons to assure the broad and consistent availability of insulin with decentralized production. Illustration by Zach Weiner of "Saturday Morning Breakfast Cereal." (http://www.smbc-comics.com/?id=935#comic)

sitting on profits collected from keeping patients addicted to consumable supplies. Some ideas:

- Jet injectors are an established technology to administer insulin that need not use and don't usually use consumable supplies.

- Radio-frequency spectroscopy of body fluid is a means of measuring blood glucose continuously that also doesn't directly need consumable supplies and has broader applicability to sensing concentrations of other components of body fluids that are of interest to medicine in general and quantified-selfers particularly.

Neither jet injection nor radio-frequency spectroscopy require any complex hardware or exotic materials to implement, so they would make good candidates for open source experimentation. Tim Cannon's recent implantation of a smartphone-sized temperature sensor in his arm (*http://www.livescience.com/40892-biohacker-tim-cannon-cyborg.html*) also shows that DIY tinkering need not remain strictly noninvasive.

Combining all these components into a working system would also result in a platform useful for taking the hypothetical open source closed-loop algorithm research to the point of real tests.

Economics of Effective Health Care and Scientific Research

The financial crisis of 2008 opened discussions on many economic topics of interest to medicine, but I have seen little public discussion taking advantage of the insights that arose in the broader economic discussions.

One issue is the kinds of incentives mentioned before in relation to consumable supplies. They need to be realigned to provide economic benefits to caregivers based on the health of the patient (the true goal) rather than on the amount or cost of services provided (which clearly should be minimized as long as the goal of patient health is achieved, both to minimize direct cost and to minimize risk of iatrogenics). In his 2008 essay in *Harper's Magazine* entitled "Our Phony Economy," Jonathan Rowe described the problem well:

> Current modes of economic measurement focus almost entirely on means...The medical system is the same. The aim should be healthy people, not the sale of more medical services and drugs. Now, however, we assess the economic contribution of the medical system on the basis of treatments rather than results. Economists see nothing wrong with this. They see no problem that the medical system is expected to produce 30 to 40 percent of new jobs over the next thirty years. "We have to spend our money on something," shrugged a Stanford economist to the New York Times. This is more insanity. Next we will be hearing about "disease-led recovery." To stimulate the economy we will have to encourage people to be sick so that the economy can be well.

The grim reality may be that we are already many decades deep into doing this —the perverse incentives have prevailed for at least that long.

Perhaps this can be addressed by aligning the interests of patients and caregivers by paying doctors only when their patients are well. This is sometimes said to have been the custom in premodern China and precolonial India. And perhaps this can be done under the structure of lodge practice, which was a common and successful structure for the provision of health care before the New Deal in America.[1]

As for the research itself, in his talk entitled "On Bureaucratic Technologies & The Future as Dreamtime," David Graeber put the problem this way: "Even though we've been pouring our research money into medicine we still don't have a cure for cancer. But we do have Ritalin, and Zoloft, and Prozac, all these things that basically make people not go completely insane despite the intense work regime they are now under." And after noting the decline in the pace of scientific discoveries after the US succeeded the UK as global hegemon and bureaucratic institutions and corporations began to dominate as a result, he stated:

> If you look at where did a lot of these discoveries actually come from in the UK, they didn't come from institutions. A lot of them came from things like rural vicars. You know, sort of eccentrics of society, they put them somewhere where they only had to do something once a week, and they would, like, study the insect life, or work on their strange theories of whatever it might be, and 90% of them were completely crazy, but 10%, that's where the patents came out of, that's where discoveries largely came out of.

Graeber then quoted astrophysicist Jonathan Katz's essay entitled, "Don't Become a Scientist," where Katz said, "It is proverbial that original ideas are the kiss of death for a proposal because they have not yet been proved to work." Graeber continued:

1. See George Rosen's history of lodge practice (*http://1.usa.gov/1jo8Anx*) and advocacy for the idea in our time at *http://bit.ly/1cWg6Oi* and *http://www.freenation.org/a/f1213.html* and an early milestone in the debate around access to medical care (*http://bit.ly/1kYXJSG*). The US tax law category that exempts these types of organizations still exists.

If you want to actually come up with an unexpected breakthrough...
you get a bunch of creative people, you give them whatever they
want, whatever resources they need, and you leave them alone for a
while. After a while you come back and most of them aren't going to
come up with anything but one or two will come up with something
you would have never imagined. If you want to make absolutely sure
that innovative breakthroughs never happen, what you do is you say,
OK, none of you guys get any resources at all unless you spend most
of your time competing with one another to convince me that you
already know what you're going to discover.

The case for putting serious money into citizen science turns out to be quite pragmatic.

Indeed, the modern research university was largely an innovation of the early modern French empire, adopted by Frederick William III's Prussia and further refined into a tool for centralizing control of the Prussian intellectual culture. The totalitarian Prussian regime needed such a tool to serve its agenda of maintaining an obedient and rigidly regimented army to hire out to other warring powers as mercenaries or, that failing, to threaten them directly. Another contemporary innovation of that regime was to require all women to register their menses with the police so that the population's fertility could be optimized, a fact noted by John Taylor Gatto in his *Underground History of American Education*.[2] The Americans who imported the research university system to America looked explicitly to such precedents as inspirational successes.[3] With our knowledge of what disastrous results these institutions ultimately contributed to in the German state that succeeded Prussia, we should be able to set a better course for ourselves than the people who committed these errors.

Decision Making Under Uncertainty and the Foundations of the Applied Scientific Method

It turns out that the statistical methods used to evaluate scientific research were born out of many of the same authoritarian, high-modern impulses (to use the terminology of Yale anthropologist James C. Scott) that drove the bureaucratization

---

[2]. Gatto, John Taylor. *The Underground History of American Education* (*http://www.johntaylorgatto.com/chapters/7b.htm*), New York: The Oxford Village Press, 2003. Chapter 7b.

[3]. 'ibid., chapter 7d. (*http://www.johntaylorgatto.com/chapters/7d.htm*)

and corporatization of the economy. The two people who contributed the most to the currently dominant way of statistically testing hypotheses, Ronald Fisher and Karl Pearson, were ardent eugenicists, and though it is only a circumstantial observation, it is worth noting that eugenics' pretenses of rationality are based entirely on selective and specious interpretations of data, on grossly oversimplified and unjustifiedly optimistic beliefs about the effects of destructive interventions, and on mistaking correlation for causation as a matter of course. Even so, Fisher warned that his test of significance should not be used to come to scientific conclusions, but rather as a tool to guide the intuition and as a possible first step before formulating a more rigorous analysis specific to the problem at hand and more sensitive to ambiguities inherent in real situations. Fisher feuded with proponents of a rival methodology, Neyman and a younger Pearson, about the correct way to draw statistical inferences until the end of his life in 1962. A hybrid of the two methods that none of them intended became the version generally taught and used. It came under criticism from Bayesian and other perspectives, and the controversy over this methodological question and many others continues in journals in several fields alongside the results supposedly vetted by the controversial methods.[4]

More recently, in 2005, John Ioannidis of Stanford surveyed the situation and explained "Why Most Published Research Findings Are False" in his publication of the same name, considering medicine specifically, commenting that, "for many current scientific fields, claimed research findings may often be simply accurate measures of the prevailing bias." Starting around 2010, Stephanie Seneff of MIT found that statin treatment for high cholesterol may be counterproductive overall and may include extreme iatrogenic risks and identified fallacious reasoning and failures to reproduce in many key research findings in favor of treatment.[5] However, Lipitor, a statin, "is the world's all-time biggest selling prescription medicine with cumulative sales topping $130 billion," and about 20 million Americans are taking statins now.[6] In 2012, Samuel Arbesman studied the turnover in scientific beliefs and cited research by Poynard et. al. finding that the half-life of a fact in the fields cirrhosis and hepatitis is about 45 years, and therefore about half of the ostensible

---

4. See Krantz, David H. "The Null Hypothesis Testing Controversy in Psychology," *Journal of the American Statistical Association*, 94, 448, Dec 1999, ABI/INFORM Global, 1372, and the class notes prepared by R. Chris Fraley (*http://www.uic.edu/classes/psych/psych548/fraley/*).

5. See Seneff's materials at her website (*http://people.csail.mit.edu/seneff/*).

6. Haiken, Melanie. "The Latest Statin Scare: Are You At Risk?" (*http://onforb.es/LHkxGk*) *http://forbes.com*, February 29, 2012.

facts a typical adult could have supposedly known about those fields during the time studied were false. [7] In a 2013 draft, Spyros Makridakis of INSEAD found no benefit, but only risk, in most cases of treatment of hypertension according to prevailing medical guidelines, and no evident overall benefit to life expectancy in treatment across the whole population.[8]

The litany continues in the introduction to a recent issue of *The Economist* dedicated to "How Science Goes Wrong":[9]

> *A rule of thumb among biotechnology venture-capitalists is that half of published research cannot be replicated. Even that may be optimistic. Last year researchers at one biotech firm, Amgen, found they could reproduce just six of 53 "landmark" studies in cancer research. Earlier, a group at Bayer, a drug company, managed to repeat just a quarter of 67 similarly important papers. A leading computer scientist frets that three-quarters of papers in his subfield are bunk. In 2000-10 roughly 80,000 patients took part in clinical trials based on research that was later retracted because of mistakes or improprieties.*

Meanwhile, Bart Kosko compares the dominant use of probability models that accommodate easy analytical manipulation, and more broadly, reliance on probability theory itself rather than broader classes of models of uncertainty, to the proverbial drunk who searches beneath the light of the street lamp for lost keys rather than where they were last seen, off in the dark.[10]

From all this, I conclude that we have nothing like a reliable mechanism for drawing useful conclusions from data in any but the simplest of situations, certainly not in those encountered in biology or medicine, and that many mistakes are hidden and frauds are perpetrated by use of distracting and impressive but meaningless rituals. I expect that someday, historians of science, and perhaps any student of the history of our times, will be aghast at all the waste and suffering we have inflicted

7. Bailey, Ronald. "Half of the Facts You Know Are Probably Wrong." (*http://bit.ly/LVisrd*) *http://reason.com*, 02 October 2012.

8. Makridakis, Spyros. "Hypertension: The Evidence." (*http://www.fooledbyrandomness.com/makridakis.pdf*) 2013 draft.

9. *http://econ.st/1hMiPix*

10. See "Lamplight Probabilities" (*http://edge.org/response-detail/23856*) and "The Sample Mean." (*http://edge.org/response-detail/10095*)

upon ourselves in this way. The question of what mechanism should be used, if even one should be used, to generate useful scientific results merits much more attention than the negligible amount it currently gets. (But it has a fair amount of mine already.)

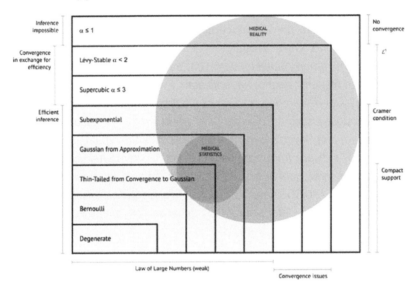

Figure 6-2. Many applications of statistics in medicine build in assumptions of unrealistically well-behaved uncertainty in the hypotheses formulated for null hypothesis testing and in methods like ANOVA. However, we know both theoretically and empirically that whenever there are inter-relationships among the entities being studied and those interrelationships are structured as usually seen in economic and social networks, the related quantities are distributed with heavy tails, moving the reality underlying the data into the categories of the outer regions of the diagram and making attempts to fit models to data difficult or impossible. Good experimental design is meant to isolate the subjects of study from these kinds of influences or control for them when they remain, but the same issues of needing things to average out arise both in measuring variability and attempting to control for it when it is undesired. Genetic, economic, and social realities may be more confounding than we currently admit, and account for some of the difficulty in reproducing research findings, even if the biological phenomena themselves truly are well-behaved despite their being based in complex nonlinear systems possessing feedback and qualities of computational character that certainly don't average out on the whole. The region labeled "medical reality" could be even larger; in a sense it is impossible to bound. Diagram adapted from Nassim Taleb's diagram (http://fooledbyrandomness.com/FatTails.html) by David Barthwell of Verge Graphics (http://vergegraphics.com/).

Nassim Taleb offers the core of a respectable answer in his advocacy of what he calls, in the terms of his pragmatic philosophy, convex tinkering. He has built the most recent phase of his career on promoting his pragmatism and the phase before that on predicting and profiting personally from predicting the financial collapse of 2008, but the need for his insights might be most urgent in medicine. He stresses the limits to knowledge in drawing distinctions between the knowable and the unknowable that are unaccounted for by the most commonly applied statistical techniques, which assume away such dangers a priori, and even usually book unmodeled variation towards the confirmation of the hypothesis under consideration. From there, he advocates all feasible control over the consequences one derives from unknowable events, making them beneficial if possible and avoiding them otherwise. In research, this amounts to much broad tinkering under conditions where successes can be noticed and turned to benefit but failures lead to no great loss, rather than high-overhead, high-stakes, overwrought, top-down research agendas. Science and medicine deserve to be practiced accordingly, as it turns out they consistently have been in their most successful embodiments since ancient times. As Graeber's observations suggest, citizen science will do so where institutional science has failed to do so.

Anthony Di Franco works at the intersections of complex adaptive systems, economics, and computing. Most recently, he designed and implemented the patent-pending accounting and transaction-planning system for decentralized finance of Credibles (https:// credibles.org/), a crowd-finance platform that hosted Oakland's People's Community Market's direct public offering earlier this year. He is a board member of the East Bay biology hackerspace, Counter Culture Labs (https://counterculturelabs.org/). He is also the author of the transgressive historical wuxia manhua, Three Sovereigns (http://3sover eigns.com/).

Big Things You Need to Know About Garage Lab Safety

Raymond McCauley

Big Thing #1: Keep It Legal

CONFLICTING IMAGES

Ask someone the first thing that comes to mind when considering biohacking and accidents, and you'll get impressions that fall into roughly two groups. One group, from the uninitiated public, might sound like a summer movie thriller all about killer viruses cut and pasted together in dank basement laboratories escaping into the wild. The other group, coming from DIYbio practitioners, is more prosaic and is concerned with handling broken glassware, containing spills, and best practices for cleaning up biohazards. These conflicting images—and the different perceptions of biohacking at the root—are worth a detailed look themselves. Happily, both sets of concerns can be addressed, at least on a basic level, with good laboratory safety practices. Having a safe home lab involves many practical elements—lab location, personal protection, anticipating problems, common safety equipment, and best lab practices—but we'll start with one of the least practical but most asked about pieces: what is legal? We'll tackle some of these other issues in future editions of BioCoder.

KINDS OF LABS

Home labs are different from community labs, such as BioCurious and Genspace, and other public maker spaces where several people come together to share access to tools and ideas. And there are even other differences with commercial and academic labs. These differences mostly work in the favor of the home experimenter. Or at least they make it simpler to start a home lab:

Simplicity
> Responsibility for the home labspace generally rests with one individual, who also controls access and supervises use.

Privacy
> Home labs are private, and users don't need to worry as much about public perception, scrutiny, strict adherence to regulations often meant to cover commercial spaces and paid employees, or worries about public access and liability. Like anything, there are tradeoffs here, too. Having just one set of eyes on a problem means that a lot of solutions get missed. And not worrying about public perception or getting a pass on commercial regulations doesn't mean that they're not good ideas. Even a home lab is required to protect other people present in your home, your neighbors, and the environment.

BREAKING THE LAW

This is actually hard to do. Surprisingly, there are few formal regulations dealing with amateur labs or typical research in biotechnology. At least, this is currently true for the United States; YMMV in other countries. For instance, in Ireland, genetic engineering labs must be licensed by the Irish Environmental Protection Agency—biohacker Cathal Garvey registered his parents' house as a Class 1 lab so he could genetically manipulate microbes of "negligible risk" legally. Even in the US, there is a patchwork of regulations dependent on where you live, although this is the exception rather than the rule. In California, at the state level, human reproductive cloning is banned, with a fine of $1 million levied for violations, making it prohibitively expensive to raise a large clone army in the Golden State. (Protip: Nevada does not even prohibit the use of state funds for human cloning, borders California, and the available huge tracts of desert land may be more suitable for a secret underground lair.) Note that most of these laws have monetary fines rather than criminal penalties. A few federal regulations exist, but most of these also lack criminal teeth. They use the carrot of federal funding to enforce experimental standards or restrictions. Bans on funding for reproductive human cloning and the reversed ban on using particular cell lines for stem cell research are good examples of using funding carrots to enforce limits on research.

WHAT HAPPENS IN THE LAB STAYS IN THE LAB

The concept of biosafety levels (BSL) falls into the same general basket—it's not a formal regulation, but it carries considerable weight. BSL is a set of lab safety criteria put forth by the Department of Health and Human Services' (HHS) Centers for Disease Control (CDC). There is no regulation that requires labs to use these standards and no government authority that issues permits. But a lab that is not in compliance may be denied government funding or can face other obstacles to working with institutions.

The BSL standard is not a regulation, but a set of guidelines, voluntarily adopted by many labs to guard against the spread of potentially harmful organisms under study. Biosafety levels come in four flavors: BSL-1 through BSL-4, with each higher level building a more strict set of precautions, affecting facilities, personal protection, and procedures for doing increasingly serious work while still protecting lab personnel and the environment:

BSL-1

Work with non-disease–causing microbes (e.g., E. coli K-12), on an open lab bench, with just a lab coat and maybe gloves and lab goggles for protection. Work surfaces are sterilized. This describes the vast majority of biohacker labs.

BSL-2

Organisms posing a small threat to people or the environment are handled (e.g., *Staphylococcus aureus*), typically in a controlled access lab with enhanced protection, biosafety cabinets, and decontamination equipment available.

BSL-3

Serious work with organisms that can be deadly if inhaled (e.g., *Mycobacterium tuberculosis*). Researchers may get special immunizations, wear respirators, and enter the negative-pressure lab through airlocks.

BSL-4

Extremely serious work with dangerous, exotic organisms (e.g., Ebola), done with glove boxes or in bunny suits. Think *The Andromeda Strain* without the self-destruct switch.

There are special versions of the BSL guidelines for facilities with animals and plants, designed to protect against the spread of animal disease and plant pests, particularly ones that might affect the agricultural economy. Hollywood sometimes features a fictional BSL-5 for really terrifying, alien invasion–type biohazards, a sort of "it goes to 11" bioclassification. Don't be fooled—BSL-4 is where it tops out in the real world.

And it's just a good idea—the BSL matches sets of facilities and procedures to levels of risk in handling hazardous organisms and is almost universally useful. Lots of biohacker labs adopt some version of the BSL-1 standards. It's a good way to ensure that what happens in the lab stays in the lab. At least be familiar with the specific requirements for a BSL-1 lab and use the parts that will work for your situation. One big implication of having a BSL-1 lab: you limit yourself to not working with human bodily fluids, including spit samples and blood spots, and you can't host mammalian cell cultures or microbial cultures of unknown origin. This is because any sample could, unbeknownst to you, contain disease vectors that would require BSL-2+ precautions.

The one area of US federal law that does have some serious and specific restrictions for biologists—and carries criminal and civil penalties—is working with organisms on the Select Agents and Toxins list. These are almost all organisms that either cause serious human disease or could devastate commercial agriculture; Yersinia pestis (Black Death) bacteria, ricin toxin, and foot-and-mouth disease virus are all on the list. Working with these legally requires special permission and facilities. Again, steering clear of this list falls under the heading of "just a good idea."

GENE TINKERING

Again, possibly surprisingly, there are few laws in the US restricting the use of genetic engineering or synthetic biology to alter the DNA of living creatures. Up until now, this field has mostly been restricted to academics and commercial professionals and has depended largely on self-regulation and liability concerns to keep risk to a minimum. In a nutshell, roughly and approximately, you're within the law if you're not:

- Working with organisms or DNA on the Select Agents list
- Selling or distributing or doing environmental release of GE drugs, food, or plant pests

- Increasing the resistance or communicability of health hazards or invasive species
- Releasing harmful GE organisms into the environment (exactly what "harmful" means is open to some interpretation)

The regulatory agencies are the FDA, EPA, and USDA's Animal and Plant Health Inspection Service (APHIS). Regulation in this area is a mess, and it's quickly evolving. Documents exist ranging from the most complete NIH Guidelines to the fictionally derived Pledge for Genetic Engineers. One request: don't give authorities a reason to harshly reassess the current stance toward self-regulation established after the historic Asilomar Conference on Recombinant DNA in 1975. Again, there is some state and local variation in these laws.

A couple of examples: on a county level, it is "unlawful for any person, firm, or corporation to propagate, cultivate, raise, or grow genetically modified organisms in Mendocino County," California, ostensibly to protect the organic status of crops grown in the region. Violators have had their GE crops destroyed by the government. Somewhat ironically, Mendocino is the source of a large portion of the state's marijuana crop and is a place where authorizes are more likely to take a torch to your GE soybeans than your field of weed. Also, as a city-level example, in Walpole, Massachusetts, performing genetic engineering work requires an annual $5 permit, which is not burdensome in itself, but has specific requirements about filing lab floor plans, procedures, training, inspections, and the formation of an institutional biosafety committee (IBC).

TAKING OUT THE TRASH

What will probably have more impact on you is where you interface with your local utilities and services infrastructure. Most local governments (at the city, county, or water district level) have regulations about what's OK to dump down the sewer or throw in the trash. What is hazardous, and how do you dispose of it? The short answer is: if you'd be uncomfortable spilling it on your dining room table, don't just flush it or throw it away. Check with your local water and trash authorities about the best way to proceed. For instance, most water authorities require that residential wastewater result from basic activities like washing and not contain hazardous waste. Specific restrictions may include things like allowable range of pH, anything that can catch on fire, materials that might cause obstructions, and even the color of the water.

Hazardous chemicals are worth their own long, tedious discussion. Hazardous chemicals include substances that meet certain criteria for ignitability, corrosivity, reactivity, radioactivity, and toxicity (including anything containing heavy metals, halogenated organics, greases, and oils). Most bioresearch does not use hazardous substances (with a few exceptions, which are a good idea to shun). Ethidium bromide (often abbreviated EtBr) has been in widespread use for staining gels in DNA work and gel electrophoresis; it's thought to be a mutagen, or cancer causing, and in the right (or wrong) concentrations can require special handling and disposal. Use safer alternatives like GelGreen, GelRed, and SYBR Safe. Polyacrylamide gels are made from acrylamide, which is a potent neurotoxin. The possibility of contaminating yourself when mixing up polyacrylamide, or of there being free acrylamide in a gel, should be enough to make you want to switch to agarose gels.

Biohazardous waste is confusing. It can be hard to define, as different institutions have different definitions of what constitutes a biohazard. And it's very possible that your local trash authorities don't have a lot of experience with this and may be prone to overreact if you approach them with biohazardous waste questions. For the purposes of a home lab, doing BSL-1 work, you can minimize your risks here by treating anything that lives or grows, or anything that touches it, as a biohazard. This means that bacterial cultures, fungus, mold, and growth media as well as countertops, gloves, lab coats, and instruments need to be sterilized or disposed of as biohazardous waste. Taken to extremes, this means that spitting on the floor is a biohazard (and not just rude) and that a puppy "accident" is a biohazard. This seems a ridiculous stretch. But context matters. Spitting on the floor in your home lab should be treated as a biohazard. So should puppy poop, if the puppy is genetically engineered but hasn't been vetted for general environmental release (also, call me—I've got a killer idea for glow-in-the-dark miniature Irish wolfhounds). Practically speaking, if you sterilize biohazardous waste, then you just have waste, and that you can throw away or flush. You can sterilize with chemicals or heat.

To sterilize with heat, use an instrument called an autoclave. On the plus side, autoclaves can also be used to sterilize growth media or glassware and are appropriate for sterilizing small amounts of biowaste. On the minus side, they're moderately expensive, even on the resale market, and have their own sets of risks for operators (they're commonly described as "pressure cookers from hell").

Most home lab operators use chemicals. Soaking liquid biowaste in a fresh 10% bleach solution for at least 20 minutes before flushing the whole solution should inactivate anything biological. But don't flush gloves or pipette tips or containers. At worst, you can find a commercial waste disposal service that will pick up your bag of biowaste for around $100. Local authorities may have specific regulations about the amounts of hazardous chemicals (including flammables like alcohol) you can store in one place, about how and where certain chemicals and waste are stored, and how much and how long biohazardous waste can be stored before it's treated. Know before you grow—check with your local authorities. County or commercial waste handlers are a really good place to start.

LIABILITY

Finally, in the US, the last resort when people are hurt or property is damaged is often hiring a lawyer and suing someone. If you burn down your apartment building, it really doesn't matter if it's from a lamp you badly rewired or if you left an alcohol lamp burning—something you could have prevented burnt several homes to the ground, and you may be liable. Even more so if your synthetic bioscience project starts spreading like crabgrass. Aside from giving general advice like, "Don't let that happen," I'll just say you can minimize your risk here by making sure you are following good lab and safety practices. Even though the BSL guidelines are just guidelines, scientists are often judged on how well they follow them. And it's good to know what your home or renter's insurance covers. If you're pursuing a hobby in your garage, you may be covered. If you're an inventor and want to commercialize, or if you are actually running a small business out of your home, you will likely need a rider on that home insurance. And if you're running a business in any way connected to a lab, there's a whole other set of laws you'll need to know about for employees, safety training, and OSHA regulations.

DON'T JUST LISTEN TO ME...

Finally, if you're doing something that may affect life and limb and livelihood, don't trust some random guy on the Internet. I am not a lawyer, and I'm not offering legal advice or opinions here, but rather encouraging you to seek out the proper authorities, consult more resources, and make your own evaluations. If you really want to know more, find other biohackers. Engaging with the biohacker community is the single best way to quickly learn better practices and improve your knowledge and your abilities. If there's not an active biohacker group where you live, start one —host a meetup. At worst, connect with like-minded folks on the Internet. The DIYbio group hosts a service called "Ask a Biosafety Expert" for questions like the ones that we're making a start on answering here. Another sterling resource is your local university. They likely have an Environmental Health and Safety (EH&S) office that does some sort of safety training and sets requirements for labs on campus. A polite, short list of questions about how they handle biosafety for microbiology and molecular biology labs can yield wonders. Ask for a checklist or to sit in on training. Other good synergies can come out of working with the right people at nearby schools.

So, really don't just listen to me. Please constructively question and challenge the things listed here. For one thing, I left out more than I included. For another, anything that has to do with safety and security works best when it's out in the open, with multiple viewpoints considered and discussed. If you're just starting a home lab, and especially if you're trying to convince a spouse or roommate that it's a good idea, saying, "Well, it's mostly not illegal" may not be the best ammunition. But it is good to know, especially if it's a starting-off place for finding more resources in your community. And if this is useful, I'm hoping we can discuss more considerations for home biohacker labs, and especially more enabling and practical topics, in coming issues of BioCoder.

References

- Thompson, Bruce and Barbara Fritchman Thompson. *Illustrated Guide to Home Biology Experiments* (*http://oreil.ly/1biUYSv*). Cambridge, MA: O'Reilly Media. April 2012.
- CDC Laboratory Biosafety Level Criteria (*http://1.usa.gov/19Orrop*)
- Ask a Biosafety Expert (*http://ask.diybio.org/*)
- BioCurious Safety Rules (*http://bit.ly/1cW3URI*)

Author's Note

I am indebted to Josh Perfetto, the founding Safety Officer at BioCurious, for much knowledge about this area, and for many discussions (and sometimes arguments) about the balance between proper safety precautions and freedom of scientific inquiry. When he left active participation at Bio-Curious, many of his duties fell to me, and I now appreciate his stance more than ever. It's more fun to argue for what everyone should be allowed to do, but much more productive to find a middle way that appropriately minimizes risk. Thank you, Josh, for starting us off right.

Raymond McCauley is chair of the biotech track at Singularity University (http://singu larityu.org/). He's also cofounder and chief architect of BioCurious (http://biocuri ous.org/), the hackerspace for biotech, a not-for-profit where professional scientists, DIY-bio hobbyists, and entrepreneurs come together to design the next big thing. He was also part of the team that developed next-generation DNA sequencing at Illumina, where he worked in bioinformatics, cancer sequencing, and personal genomics. Follow him at @raymondmccauley (https://twitter.com/raymondmccauley).

Molecular Tools for Synthetic Biology in Plants

A First-Generation Open Bioinformatics Workshop

Ron Shigeta, Niranjana Nagarajan, Shriram Bharath, Wilifred Tang, Tony Hecht, Alex Alekseyenko, Bryce Wolfe, Corey Hudson, Jamey Kain, Urvish Parikh, and Scott Fay

This project was sponsored by Counter Culture Labs and Berkeley Bio Labs. Please address correspondence to *ron@berkeleybiolabs.com*.

Abstract

Synthetic biology has had profound effects on human life. It has provided more effective anti-malarial medicine, cheaper insulin, new useful biomaterials, and greener biofuels. However, much remains to be learned in order to synthesize proteins more efficiently. To explore the potential of the DIY biology movement to engage in meaningful synthetic biology bioinformatics research, we developed a bioinformatics workshop to study determinants of protein expression levels in plants. We extracted possible ribosome binding and translation initiation sequences and looked for correlations with experimentally determined protein levels using publicly available datasets for the widely studied plants *Oryza sativa* and *Arabidopsis thaliana*. The working group was open to the public and met every other week for three hours, typically starting with a short, relevant presentation followed by hands-on data work. We aim to develop, experimentally validate, and publish our consensus sequences, anticipating that our work will be useful for plant synthetic biology research. We hope our experience will be a model for future community projects that serve the dual purpose of educating curious members of the public while also generating useful scientific results.

Introduction

Advances in sequencing technology have produced an avalanche of biological data over the past 12 years. The bottleneck in discovery has consequently shifted from data generation to data analysis, suggesting that much data is not used to its full potential.[1]

Crowdsourcing is one technique used to gain more insight from existing biological data. Putting the diverse eyes and hands of the general public to the purpose of bioinformatics is not new.[2] Examples include protein[3] and RNA folding (*http://eterna.cmu.edu/web/*), and both paid Ingenuity Systems (*http://www.ingenuity.com*) and unpaid[4] curation of literature.

Rather than approach a problem strictly as professionals, we developed an open source DIY workshop where scientists and the public worked together to tackle a synthetic biology project resulting in a publishable outcome. The problem to be solved would need data from completely open sources and not require difficult analysis. A modest goal was set to do a survey of plant translation initiation motifs, aiming to create an open source parts list for controlling translation in metabolic engineering and synthetic biology. Working meetings were posted through Counter Culture Labs and Berkeley Bio Labs (groups with >100 members each) on Meetup.com and met every week or two over three months.

---

1. Lockhart, David J. and Elizabeth A. Winzeler. "Genomics, gene expression and DNA arrays," *Nature*, 405, 827–836.

2. See Good, Benjamin M. and Andrew I. Su. "Crowdsourcing for bioinformatics," *Bioinformatics* 29, 2013, 1925–1933 and Marbach, Daniel et al. "Wisdom of crowds for robust gene network inference," *Nature Methods* 9, 2012, 796–804.

3. Lane, Thomas J. et al. "To milliseconds and beyond: challenges in the simulation of protein folding," *Current Opinion in Structural Biology*, 2012.

4. Hingamp, Pascal et al. "Metagenome annotation using a distributed grid of undergraduate students," *PLOS Biology* 6, 2008, e296.

Plants offer many advantages as systems to do fine-tuned biological engineering (e.g., modification to enhance production of economically valuable terpinoid,[5] or modification of lignin biosynthesis to expedite biofuel synthesis[6]). There is a paucity of published information, however, on how to control sets of genes working in concert. Use of small sequence motifs as ribosome binding site parts for synthetic biology has been proposed in bacteria,[7] and similar parts have been produced for yeast (*http://parts.igem.org*). Estimates for RBS parts in prokaryotic systems show that the translation level of a gene can be shifted by greater than an order of magnitude, indicating their potential utility in synthetic biology projects. Generating an estimate of the regulatory power of plant translation initiation motifs was thus seen as a useful goal for our project.

In most eukaryotic plant genes, the 5' cap of the mRNA transcript acts as the ribosome binding site and the Kozak sequence acts as the signal for translation initiation. Due to the bacterial origins of the chloroplast, transcripts of genes encoded within the chloroplast genome contain distinct consensus sequences in comparison to transcripts from the nucleus. Instead of the 5' cap, there is a short motif called the Shine-Delgarno sequence where the ribosome binds and then initiates translation, generally eight nucleotides downstream, though this distance varies. Although there has been some experimental work on ribosome binding sites and Kozak sequences in plants,[8] genomic-scale surveys have not been performed.

Here we use publicly available, combined RNA- and protein-expression data for both nuclear and chloroplast genes to estimate the power of the ribosome binding and translation initiation sequence motifs to initiate translation. These are initial results; experimental confirmation of the motifs will follow.

5. Moses, Tessa et al. "Bioengineering of plant (tri) terpenoids: from metabolic engineering of plants to synthetic biology in vivo and in vitro," *New Phytologist*, 2013.

6. Li, Xu et al. "Improvement of biomass through lignin modification," *The Plant Journal* 54, 2008, 569–581.

7. Salis, Howard M. et al. "Automated design of synthetic ribosome binding sites to control protein expression," *Nature Biotechnology* 27, 2009, 946–950. See also *http://bit.ly/1iPO5Bu*.

8. Shine, J. and L. Dalgarno. "The 3'-Terminal Sequence of *Escherichia coli* 16S Ribosomal RNA: Complementarity to Nonsense Triplets and Ribosome Binding Sites," *Proceedings of the National Academy of Sciences of the United States of America* 71, 1974, 1342–1346.

Methods

PLANT GENOME SURVEY AND MOTIF EXTRACTION

A broad survey of the translation initiation motifs from both the TAIR10 Arabidopsis thaliana genome build[9] and the IRGSP 1.0 Japanese Rice Genome[10] was carried out. In order to capture translation initiation motifs as well as possible leader peptide sequences, the gene description GFF files were used to extract the 25 bases before and 18 bases after the start codon of each gene for each genome build. The terms "CDS" or "mRNA" were used to extract protein coding regions. With the data from the rice genome, we were unable to separate coding sequences in all three possible reading frames; therefore, we excluded coding sequences that did not initiate with the canonical "ATG" start codon.

CHLOROPLAST SURVEY AND MOTIF EXTRACTION

Because of the small number of genes in the chloroplast, a broad collection of motifs were also extracted for chloroplasts. The GenBank chromosome sequences were scraped from the Choloroplast DB webpage[11] and used to extract motifs using Biopython.[12] This yielded 11,810 initiation motifs from 109 organisms, which gave good consistency in the start codon with translation initiation generally occurring 8 nucleotides downstream of the ribosome binding site, as expected.

TRANSCRIPTOME DATA

As we could find no publicly available matched proteome/transcriptome datasets, we obtained arrays from arabidopsis leaf and rice leaf. All replicate arrays for

9. Lamesch, Philippe et al. "The Arabidopsis Information Resource (TAIR): gene structure and function annotation," *Nucleic Acids Research* 36, 2007, D1009–D1014.
10. Kawahara, Yoshihiro et al. "Improvement of the Oryza sativa Nipponbare reference genome using next generation sequence and optical map data," *Rice* 6, 2013, 4.
11. Cui, Liying et al. "ChloroplastDB: the Chloroplast Genome Database," *Nucleic Acids Research* 34, 2006, D692–D696.
12. Cock, Peter J. A. et al. "Biopython: freely available Python tools for computational molecular biology and bioinformatics," *Bioinformatics* 25, 2009, 1422–1423.

noontime leaf expression in adult plants were obtained from Gene Expression Omnibus,[13] via GEOSearch.[14]

The following table lists the datasets that were chosen in the workshop session.

Experiment	Array designation	Sample description
GSE11966	GSM302918	Expression data from rice leaf rep 1
GSE11966	GSM302919	Expression data from rice leaf rep 2
GSE22788	GSM563421	Rice Kitaake Leaf rep1
GSE22788	GSM563422	Rice Kitaake Leaf rep2
GSE22788	GSM563423	Rice Kitaake Leaf rep3
GSE24048	GSM591761	Control Azucena leaf biologial rep 1
GSE24048	GSM591762	Control Azucena leaf biologial rep 2
GSE24048	GSM591763	Control Azucena leaf biologial rep 3
GSE24048	GSM591764	Control Bala leaf biologial rep 1
GSE24048	GSM591765	Control Bala leaf biologial rep 2
GSE24048	GSM591766	Control Bala leaf biologial rep 3

Overall, 10 arrays, including replicates of 4 separate measurements, were downloaded as Affymetrix Arabidopsis Genome ATH1 and Rice Genome Array CEL files. These were scaled using the MAS5 algorithm[15] using the *affy* Biocoductor library[16] in R.[17] The resulting dataframe was reduced to mean, median, and standard deviation estimates for each probe set for Rice and Arabidopsis. The results were that the mean measurement standard deviation was 71% and the median differed

13. Barrett, Tanya et al. "NCBI GEO: archive for functional genomics data sets–update," *Nucleic Acids Research* 41, 21 D991–D995 (2012).

14. Zhu, Yuelin et al. "GEOmetadb: powerful alternative search engine for the Gene Expression Omnibus," *Bioinformatics* 24, 2008, 2798–2800.

15. Wei Keat Lim, Wei Keat et al. "Comparative analysis of microarray normalization procedures: effects on reverse engineering gene networks," *Bioinformatics* 23, 2007, i282–i288.

16. Gautier, Laurent et al. "affy—analysis of Affymetrix GeneChip data at the probe level," *Bioinformatics* 20, 2004, 307–315.

17. R Deveopment Core Team. "R: A Language and Environment for Statistical Computing." Vienna, Austria: the R Foundation for Statistical Computing, 2013.

by 19% from the mean, indicating a reasonable sample variance that was satisfactory where doubling of intensities is considered significant.

PROTEOME DATA

The Rice Proteome Project has a comprehensive set of quantitative proteome estimates from 2D SDS PAGE gel including different stages of the plant growth and portions of the plant as well as an organelle survey. Quantitation from gel densitometry, MASCOT scores, and UniProt associations were downloaded as tables.[18]

Only a few hundred measurements were found from multiple sources for Arabidopsis, which did not cover the organelles explicitly and less than 10% of the known leaf proteome. As the data proved to be inadequate for this study, the Arabidopsis survey had to be set aside.

UniProt identifiers were mapped to rice probe set identifiers. Many of the UniProt identifiers were directly mappable to probe set (236 out of 554 identifiers, split among 123 chloroplast, 235 mature leaf, and 196 seedling leaf probes) using the Rice Coexpression Database.[19] The remaining UniProt identifiers were manually mapped, BLASTP searching UniProt sequence against the Oryza sativa Nipponbare reference genome.[20]

TRANSLATION INITIATION ESTIMATION FOR MOTIFS

The interrelationship between Rice Gene, Protein in the Proteome set, and the MicroArray Probe set required several data sources. The Probe Set Annotation data for the Rice IVT Expression Array was extracted from the Probe Set Annotation CSV file provided by Affymetrix.[21] Because UniProt accessions drift over time, the Rice Proteome, which was generated circa 2004, had no protein accessions that were in UniProt. Data relationships to gene names and probe sets were assembled through multiple processes. Reviewing archival Rice Genome Array annotations, we were able to find about 50% of the probe set mappings we needed, and the rest

18. Tanaka, N. et al. "Proteomics of the rice cell: systematic identification of the protein populations in subcellular compartments," *Molecular Genetics and Genomics* 271, 2004, 566–576.

19. Sato,et al. "RiceXPro Version 3.0: expanding the informatics resource for rice transcriptome," *Nucleic Acids Research* 41, 2013, D1206–D1213.

20. Kawahara et al. "Improvement of the Oryza sativa Nipponbare reference genome using next generation sequence and optical map data," *Rice* 6, 2013.

21. Liu et al. "NetAffx: Affymetrix probesets and annotations," *Nucleic Acids Research* 31, 2003, 82–86.

were recovered manually by searches of *http://uniprot.org* and if necessary, BLAST alignment of nucleotide sequences against the Rice Genome at *http://msu.edu.*

For genes that had proteome protein concentration estimates, the translational coefficient, Θ, for a given gene was estimated as the ratio of the protein to the mean RNA concentration as estimated by the microarray intensity.

In order to reduce the influence of outliers, the mRNA concentration was taken as proportional to the median of the microarray values from the 10 datasets. This reduced the range of the values by two logs compared to taking the mean microarray probe set intensity.

Results

GENOME SURVEYS

Surveys of the nuclear chromosomes of *Arabidopsis thaliana* and *Oryza sativa japonica* yielded thousands of sequence motifs. A conventional logo survey[22] shows the expected Kozak sequence in the nuclear genes (see Figure 8-1). In the case of chloroplast chromosome, since the Shine-Delgarno sequence does not have a fixed location with respect to the start codon,[23] the weblogo does not show any appreciable signal (see Figure 8-2).

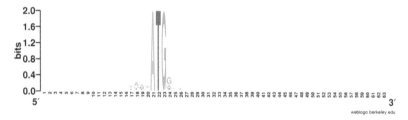

weblogo.berkeley.edu

Figure 8-1. Sequence logo of chromosome 1 of Oryza sativa japonica, derived from 2,134 sequences, restricted to those initiating with an ATG codon. This logo shows a canonical Kozak motif surrounding the initiating ATG. The x-axis represents the nucleotide position 20 bases upstream and 20 bases downstream of the ATG initiation codon. Some information in the wobble bases (third position) shows in the coding portion of the sequence. The other chromosomes were similar.

22. Crooks, Gavin E. "WebLogo: A Sequence Logo Generator," Genome Research 14, 2004, 1188–1190.

23. Hirose, Tetsuro and Masahiro Sugiura. "Functional Shine-Dalgarno-Like Sequences for Translational Initiation of Chloroplast mRNAs," Plant and Cell Physiology 45, 2004, 114–117.

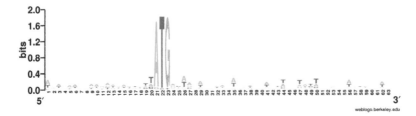

Figure 8-2. Sequence logo of chloroplast translation initiation motifs. The x-axis represents the nucleotide position 20 bases upstream and 20 bases downstream of the ATG initiation codon. Bias in the wobble base of the codons is much more pronounced in this logo since only 81 sequences were available to analyze in Arabidopsis chloroplasts.

TRANSLATION INITIATION ESTIMATES

The relative power estimates for the proteome-to-transcript ratio range over 12 powers of natural log (see Figure 8-3), which is 165,000. The average value is –2.4 with an asymetrical distribution, with a greater range for enhancements to protein production ($\Theta > 1$).

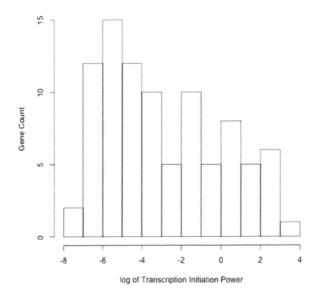

Figure 8-3. Histogram of relative protein to mRNA abundance ratio (power) for the chloroplast proteome. Using the median value of the microarray intensity, the power varied over a factor of 58,000.

The correlation between mRNA and protein available in the cell turned out to be poor—the mRNA and proteome scores had a correlation of 0.12, which implies that there are likely several factors that are influencing both of these numbers that go into Θ, indicating that the model is too simple.

NEXT STEPS

Though the workshop has performed some novel analyses, this is a preliminary work. It's clear the estimate of transcript initiation has a tremendous amount of uncertainty associated with it. Microarray probe sets are not distinctly comparable with each other, as the specific sequences of the probes vary in their target affinity.

An abundance of cell processes can affect the actual amount of protein produced compared to the mRNA reported by a microarray. Just a few of these may include nonsense mediated decay, inhibitory RNA, post-translational editing, protein sorting among cellular compartments, and secondary structure in the mRNA.

Still for the largest and smallest Θ values, the values determined might give some correlation with strong and weak translation. We will next test the leader sequences associated with the largest transcription initiation power in vivo in collaboration with the Glowing Plant project. To this end, we'll be taking the motifs for the 10 largest and some smaller Θ motifs and installing the sequence into a plasmid that can be validated in a plant cell by quantitation of florescence from a GFP versus a control construct with its current constitutive motif sequence. When their relative strengths have been determined, the parts themselves will be placed in the GoldenBraid public repository.[24]

The collection of motifs will also enable us to examine chloroplast Shine-Delgarno sequences and their relative effects on translation.

OPEN WORKSHOP

One of us (Ron Shigeta) initiated the open workshop as an experiment to bring together the populations of curious laymen, experienced wet biologists, and software engineering talent in the East Bay area, and all three of these groups were represented in the attendees. In addition, several working bioinformaticians contributed.

The project was structured to give an introduction and purpose to looking at a variety of publicly available biological data. The first five meetings each were spent

---

24. Sarrion-Perdigones, Alejandro et al. "GoldenBraid 2.0: A Comprehensive DNA Assembly Framework for Plant Synthetic Biology," *Plant Physiology* 162, 2013, 1618–1631.

on a category of biological data: chromosomal sequences, individual open reading frames, microarray data, quantitative proteomics data in 2D polyacrylamide gel electrophoresis, and quantitative proteome data from gas chromatography/mass spectroscopy. In each of these sessions, data was gathered from public sources, and participants had hands-on experience with the raw data. Attendance ranged from 25 to 30 participants. As a public-scientific–interface event, hands-on work with data and computers was quite engaging, and several useful scripts were written to process the data in Python and R.

The following two months of biweekly data analysis sessions were less fully attended, with an average of two to four participants. Possible reasons for this decline include lack of understanding of the subject matter or technical skills needed to fully participate, inability to commit to an extended project, and unclear direction or incentives to continue. The more open-ended nature of data interpretation and analysis is also a difficult process to relate to an introductory course; it was difficult for newcomers to biology to attach to these tasks.

Future workshops may be structured into beginner, intermediate, and advanced levels that would be more accessible to participants from diverse educational backgrounds and will likely be shorter in length to reduce attrition. Another idea is to take on a project with the sole goal of doing that project, rather than anticipating publishable results. As an experiment, the workshop did succeed in bringing together a range of talent and covered a broad set of biological data.

Slides for these sessions are available at *http://boundaryconditions.org/biology.html*. When we have completed screening out parts, scripts, data collected, and analysis for this project, it will be made available at *https://bitbucket.org/ronbo/glowingplantparts*.

Acknowledgments

The authors would like to thank SudoRoom, a tech makerspace in downtown Oakland, California, for physically hosting the workshop. We would also like to thank the many other individuals who came to the workshop at one time or another: Felicia Betancourt, Ryan Bethencourt, Jack Cunha, Cristina Deptula, A. Dangerfield, Brian Gordon, Carl Gorringe, Rajat Jain, Ahnon Milman, and Heather Wilson.

Biological Games
Pac-Man of the Microscopic World

Keith Comito

If you have ever played fetch with a dog or hide-and-seek with a particularly crafty cat, then you have played a biological game. In fact, humanity has a long and morally complex history of games and rituals involving other organisms, from the bull leapers of the Minoan civilization to the chariot races of ancient Rome. Now, through the application of biotechnology, such games are beginning to extend into the microscopic world, illuminating new opportunities for entertainment, education, and the understanding of ourselves as ethical and social creatures within the larger community of life.

It was about a year ago that I first heard the term "biotic games" during a workshop at Genspace (*http://www.genspace.org*) exploring the research of Stanford's Dr. Ingmar Riedel-Kruse (*http://www.stanford.edu/group/riedel-kruse/research/biotic_games.html*), in which unicellular paramecium are manipulated via electric currents in order to play Atari-style games such as Pac-Man and Pong. Dr. Ellen Jorgensen and computer engineer Geva Patz demonstrated this behavioral response to electricity, known as galvanotaxis, by placing paramecia in an area bounded by four electrodes—as one of the electrodes became negatively charged, the paramecia swam in its direction, only to make a u-turn as this electrode became positive and another negative. This immediately sparked in my mind the wonders of the 1982 landmark film *Tron* (which, in turn, was inspired by Pong): beings in the machine, tiny organisms living out their own lives, yet contributing to a grander process.

Having spent many years developing games and interactive media, both for computers and mobile devices, the potential impact of this technology became quickly apparent to me—not just in gameplay, but also in terms of possibilities for living, interactive art projects as well as interesting social commentary. Much has been said, for instance, about the relationship between the player of a game and his avatar within the game: do you control the avatar, or do you become the avatar? What does it mean if your avatar is actually another living organism? The incorporation of life into gaming also has the potential to promote thoughtful reflection about humanity's relationship to nature and the prospect of interspecies cooperation, working to combat the stigma of dehumanization and desensitization commonly attributed to video games in general. Other members at Genspace were of similar mind, and together, Dr. Oliver Medvedik, cofounder of Genspace, Sarah Choukah, Genspace member and PhD communications student, and myself decided to pool our respective expertise in biology, microcontrollers, and game development to take this technology to the next level.

The first design decision we made was to use an iPhone as the game platform (see Figure 9-1), unlike Dr. Riedel-Kruse's original model, which relied on a desktop-based Adobe Flash interface. We made this choice both because we wanted the unit to be portable, and because we felt that having the game run on such a ubiquitous device would make the underlying technology more real to the populace, rather than just a scientific curiosity. This approach was not without its drawbacks, however, as the intense graphical processing required to motion-track multiple paramecia is difficult to pull off on relatively limited mobile platforms. While I was thus busy trying to wrangle the iPhone to this purpose, Oliver and Sarah had their own hands full with engineering the control mechanism. Complicating our respective issues was the fact that the 2013 World Science Festival at Innovation Square (*http://worldsciencefestival.com/events/innovation_square*) was set to begin just one week after we had embarked on our project, and we really wanted to present.

Figure 9-1. iPhone with microscope attachment positioned above paramecia stage

Sometimes science is conducted in an orderly manner, with lab coats and latex gloves, and sometimes science is conducted frantically into the wee hours over half-eaten boxes of pizza and flashing LED lights. This was the latter. But it had worked —after a solid trifecta of all-nighters, we possessed an operational prototype. Using barely more than electrical contacts and assorted Lego blocks, Oliver and Sarah had created a basic joystick, which was then connected to an Arduino microcontroller (see Figure 9-2). The Arduino, in turn, was connected to the electrodes bounding the paramecia stage, controlling them based upon the input of the joystick. Above all of this, on a platform constructed of equal parts wood and rubber bands, hovered the iPhone, with its camera augmented by an affordable microscope attachment and trained upon the paramecia stage. As the paramecia moved, my code employed a careful combination of graphical filters to track their motion, and based upon this data generated a game similar to Pac-Man, in which the microorganisms collected Tron-like discs. There was still a kink or two to work out, such as the contacts between the Arduino and the paramecia stage becoming loose and impairing joystick control, but it was nothing we couldn't smooth out, we thought, during setup at Innovation Square after a good night's rest.

Figure 9-2. Arduino code controls voltage of the electrodes based on input of the joystick

Oliver has described what followed as performing a magic trick for the first time live and hoping you don't saw an audience member in half, which I find both hilarious and entirely accurate. In addition to our little joystick problem being not so little (thanks in part to the throng of grasping children's hands), the elements at the outdoor festival had conspired against us—the June heat was causing the iPhone to overheat as well as threatening to evaporate the liquid in which the paramecia were swimming, forcing Oliver and Sarah to perform continual water bottle and ice pack triage. Additionally, strong gusts of wind were causing noise in the motion detection, requiring me to tweak the code constantly in real time to keep things running smoothly. All this combined with fielding hundreds of questions, as well as managing dozens of paramecia-controlling gamers, makes for one hectic day. In the end, however, it all came together. Adults and children alike were fascinated by the concept of playing with microorganisms, and many even suggested that we launch a Kickstarter so they could buy such a system for home use. They praised the retro-graphical style and chiptune music created for the game and gave us myriad suggestions for improvements such as a central electrode to gather paramecia together—illustrating how even children can contribute to fields such as biotechnology if given the chance. In short, the response was incredible; our magic trick had worked (*http://bit.ly/1j0ag07*).

Emboldened by the successful proof of concept, we have continued to work on our prototype. By the time we presented at NYC Resistor's (*http://www.nycresistor.com*) 2013 Interactive Show, for instance, we had laser-cut a clear acrylic housing, solving the wind problem while still allowing viewers to glimpse the system's internal workings (see Figure 9-3). The knowledgeable hosts and audience members at this event even helped us solve our heat problem on the spot with an appropriately sized heat sink, giving just one example of the utility of atelier-style hacker spaces such as Resistor and of the Maker movement in general. At the most recent event we attended, Maker Faire 2013 (*http://makerfaire.com/new-york-2013/*), we showcased the addition of an arcade-style tilt feature, which prevents an accidental (or intentional) bump of the device from skewing the game in the player's favor. What has remained constant throughout every venue, however, is the overwhelming interest and enthusiasm people have expressed regarding this technology. This paves the way for our team and others to move forward with new and interesting applications —massively multiplayer online games in which people communicate with one another via other living creatures, self-reflexive systems in which organisms control themselves in ways they never have before, generative musical applications in which the composers are single cells...

Figure 9-3. Arduino, paramecia stage, and heat sink within clear acrylic housing

It is not just in the realm of entertainment that this technology has application, but also in provoking thoughtful dialogue regarding ethics, free will, and what it means to be human among the various forms of life that inhabit our world. This was clearly driven home to me when I was asked by a young girl if it hurt to control the paramecia, and if their movement was choice or compulsion. To her first question, I was happy to answer no, as paramecia do not have nervous systems, but as for her second: that answer is not so simple. Would not the traffic flows of human beings to a city in the morning and away from it in the evening look like a compulsion, too, from a suitable distance? Even if a paremecium's galvanotaxi s nothing but a consequence of ion channels reacting to voltage gradients, is this really that different, in terms of free will, than electrical impulses traveling throughout a brain?

It is my hope that our game and others like it will go beyond even these questions and invite people to consider a future where we find a way to cooperate with organisms of every scale. Such an approach, characterized by partnership with—rather than subjugation and destruction of—nature has the potential to foster the sustainable future all of us want and can achieve.

@keithcomito (https://twitter.com/keithcomito) is a mathematician and computer programmer currently based in New York City. In addition to developing mobile applications, he explores the intersection of technology and biology at the Brooklyn community lab Genspace. Keith also studies martial arts, writes music, and hosts a retro video game–themed YouTube channel.

Mission Possible:
Ghost Heart Protocol

*Patrik D'haeseleer, Matthew Harbowy, Ahnon Milham, and
Maria Chavez*

This year for Halloween, we decellularized a pig heart. We got a nice bleeding heart,
hooked it up to some plumbing, stripped out all its cells with enzymes and detergents, and then bottled the thing in Everclear. Because every mad scientist's den
deserves to have some mad science on display.

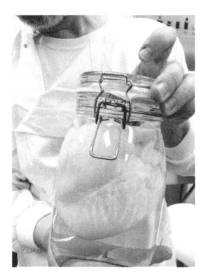

Figure 10-1. Ghost heart in a jar

For those of you interested in trying
this yourself, we've posted a detailed Instructable (*http://www.instructables.com/id/
Ghost-Heart-in-a-Jar/*). Don't be put off by
the complexity of the procedure—when it
comes down to it, the hardest part of this
project is the plumbing. Actually, this
would be a great project for a parent to do
with a kid who's really into bioscience and
wants to go a little farther than the dissec-
tions in AP biology at school.

So why would you want to decellular-
ize a heart? This is a technique being developed that may eventually produce organs for transplantation, composed of a
patient's own cells (*http://bit.ly/1joakgw*).
The idea is to take a donor organ and strip
it of all its cells, leaving nothing but the extracellular matrix that held the cells in
place. This scaffold of connective tissue—called a "ghost organ" for its pale and
almost translucent appearance—can then be reseeded with a patient's own cells,

with the goal of regenerating an organ that can be transplanted into the patient without fear of tissue rejection.

Amazingly, mouse and rat ghost hearts reseeded with cardiac precursor cells will actually start beating autonomously (great videos here (*http://bit.ly/KI5Ge8*) and here (*http://bit.ly/LVkasC*)). Some of those techniques seem daunting, but the Pelling Lab (*http://www.pellinglab.net/tag/decellularization/*) at the University of Ottawa has already illustrated how easy and eminently DIY-able the decellularization process is; in the case of a mouse heart (*http://www.pellinglab.net/decellularization/*) or even of a whole steak (*http://www.pellinglab.net/was-steak/*), it could be as simple as soaking in detergent for a while.

Having spent a good amount of time thinking about related tissue engineering techniques for the BioPrinting community project at BioCurious, it seemed only natural to try our hand at decellularization as a Halloween project. So last year, half a dozen strong-stomached biohackers got an assortment of chicken hearts and gizzards from the grocery store and made an impromptu run at decellularization. Every participant got a chicken heart to play with and tried a unique combination of detergent choice, number of distilled water and saline rinses, etc. In the end, we wound up with several jars of chicken heart in detergent soup (*http://www.meet up.com/BioCurious/photos/11068182/*) on the shake platform. Shaking, and shaking, and shaking, for days on end with very little effect. Sure, we had a few hearts that were getting noticeably paler, but the whole mess was definitely getting noticeably stinkier as well! Lessons learned. We decided to table the experiment for the year and try again for Halloween 2013. Next time, we plotted, we would use a (more human-like) pig heart, and perfuse detergents through the organ, to make sure all the cells in the tissue are broken up and flushed out equally.

Fast forward one year: Cameron at BioCurious managed to get us some pig hearts from his sister's farm, so we planned to do a test run at Science Hack Day San Francisco (*http://sf.sciencehackday.com/*). Unfortunately, hearts that have been butchered for food typically have the arteries at the top of the heart cut off and a slash through the chambers of the heart to let the blood drain out, which makes them entirely unsuited for our perfusion experiment. Instead, we cut some slices from the heart and tested a range of digestive enzymes on them (*http://bit.ly/ 1kYZwan*).

What we needed was a pig heart butchered specifically for biological experiments (called a "bioheart" in the trade). So we got ourselves some bloody pig biohearts from a local meat wholesaler, ordered some simple reagents, located a pump (thanks, Eric!), got some plumbing supplies from Home Depot, assembled a motley crew of coconspirators from Counter Culture Labs (*http://www.counterculture labs.org/*) and BioCurious, and decided to try again. With mere days before Halloween, we descended on BioCurious for a full-day experiment.

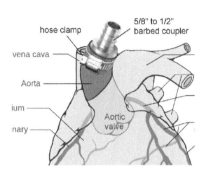

Figure 10-2. Some assembly required: pig heart with hose clamp and barbed coupler attached to the aorta, ready for retrograde coronary perfusion

The published protocols take 10–13 hours to complete, but since it took us hours to get everything setup correctly, we ended up adjusting the durations on the fly to fit into the time remaining. And since we were just making a showpiece rather than a ghost heart that could be successfully reseeded with stem cells, we made some judicious substitutions in the reagents: using plain tap water instead of gallons and gallons of reagent-grade water, kitchen salt in tap water instead of phosphate buffered saline, OxyClean (sodium percarbonate) instead of peracetic acid, and 151-proof Everclear instead of reagent-grade ethanol. If you wanted to do a real hardcore DIY version, you should even be able to replace the initial trypsin perfusion with digestive enzymes from the health food store (bromelain seemed to work quite well) and replace the lab-grade detergent with a concentrated shampoo! We didn't have access to a high-volume peristaltic pump, but a buffer recirculation pump seemed to work OK. Again, you could completely DIY this part by using a fountain pump from the hardware store, or even just an elevated bucket (six-foot elevation should give you about the right fluid pressure).

Figure 10-3. Residual blood leaking out of the heart during the initial perfusion steps, as we push fluids through the heart muscle

One nice aspect about this experiment is that it uses fairly mild reagents, since the end result should be able to support cellular growth. But you're also working with lots of fluids under pressure, and you really, really don't want assorted pig heart juices squirting into your eyes. So make sure to wear some goggles at least. Also, if you're using a fountain pump, keep in mind that they're typically immersion pumps and not designed to handle conductive saline solutions. So try at your own risk...

This was probably one of the most ambitious single-day experiments we have been involved in—especially since none of us had any experience in doing this kind of thing. We frankly surprised ourselves with how well it worked (including some successful emergency cardiac surgery using super glue!), and we ended up with a stunning showpiece. Although initially daunting, the techniques are well within reach of the DIY community and can be used to educate and begin discussions on the latest in organ engineering research.

Check out our Ghost Heart in a Jar Instructable (*http://www.instructables.com/id/Ghost-Heart-in-a-Jar/*) for all the details. After all, Valentine's Day is just around the corner.

Patrik D'haeseleer is a bioinformatician by day, mad scientist by night. He is a cofounder of Counter Culture Labs, community projects coordinator at BioCurious, and scientific advisor of the Glowing Plant project, none of which are in any way related to or funded by his day job at the Lawrence Livermore National Laboratory. The views presented here are his own and do not represent those of BioCurious, the Glowing Plant project, or LLNL.

Matt Harbowy is a scientist and hacker. He serves on the board of Counter Culture Labs.

Ahnon Milham is a citizen scientist and cofounder of Counter Culture Labs. The views presented here are her own, and she is pleased that they do indeed represent those of Counter Culture Labs as well.

Maria Chavez is a marketing professional with a background in medical research. She volunteers as Class Manager at BioCurious. The views presented here are her own and do not represent BioCurious.

How We Crowd-funded $484k to Make Glowing Plants

Antony Evans

One of the goals of the Glowing Plant project was to inspire others to look at crowd-funding as a way to reach their own goals within DIYbio/synthetic biology. We've been pleased to see a number of bioengineering campaigns launch; however, they have been unable to raise similar amounts of funding. Therefore, we want to share our tips and tricks in hopes that they will be useful to others embarking on a crowdsourced funding journey.

Preparation

Once a crowdfunding campaign gets underway, you will become overwhelmed and insanely busy. Also, crowdfunding campaigns have a significant momentum effect, so it's important that they are as well presented and developed as possible at launch. Planning for the campaign is critical.

We started formal planning eight months before the launch date. The most important thing at this early stage was discussing our plans to engineer a glowing plant with everyone we knew and met in person, including at meetups and other events.

Sharing the project openly and transparently up front achieved three goals:

1. We wanted to know if people were excited by and wanted to talk about the project. From our discussions, we learned our supporters wanted the engineered seeds.

2. It established a personal relationship with people, so when they saw the project, they felt a personal connection to it.

3. It allowed us to build partnerships broadening our social network, as the reach of the campaign is proportional to the size of the network. Many of these partners also became part of the project and helped in other ways.

We also did research online, looking at other campaigns (backing a few we liked) and seeing what did and didn't work. Both Kickstarter (*http://www.kickstarter.com/help/school*) and Indiegogo (*http://www.indiegogo.com/crowdfunding-tips*) publish information helping project creators plan their projects, and we took note of what they deemed important as well as what other people online thought (*http://bit.ly/1aJOxe9*).

As part of this research, we learned that 80% of projects that get to 20% of their funding succeed, so this became our first target milestone. We realized that getting to 20% would be easier if we closed a few big backers early on, so we went on the road and sold those rewards in person, sometimes giving them additional perks as part of the package. One of these was a $10k backer, Cambrian Genomics, which supported the project within the first hour. This gave us instant credibility that we could reach the target, igniting a wave of optimism online.

Another tip we got was from the founder of Pebble (*http://kck.st/1c4noDV*), who gave a talk at Singularity University. He convinced us to look at the landing page and treat it like a product in itself. You don't launch a product without getting user input (I've made that mistake before!). We showed the preview page to over 100 people in the month prior to launch. These people included our advisors, roommates, friendly mailing lists, and even a few people I stopped on the street in San Francisco.

The feedback we got from this group (e.g., your rewards are too expensive, the landing page is too complicated, and the video could be more exciting) was incredibly valuable. This was an iterative process; as we made the changes, we sought more opinions.

We also received landing page feedback from the Kickstarter product design category manager. We got criticism from scientists about this after launch, but it was the right decision. The landing page should be simple and most importantly actionable (back the project today!). That's why we stripped away all the text and just used images on the page. Remember, most backers don't understand the details and trust you to know them yourselves. Another reason for not going deep into technical details is that people will start discussing on social media platforms how you are going to do the science, which drives traffic and maybe some new ideas you hadn't considered.

Clarity around regulation is incredibly important, because backers want product, and they want to know what you need to do to get it to them, as well as the project-execution risks. You have to reach out to the relevant regulatory agencies (USDA, EPA, or FDA) before launching the campaign. We found all of these agencies keen to engage early and approachable and frank in their opinions, especially if you get a personal introduction.

One campaign we studied intensely was "The Ten Year Hoodie." (*http://kck.st/KHLLwC*) We were amazed at how much money they raised for such a simple product. We were really inspired by the story arc and energy of their video. What really struck us, though, was how the campaign was about more than just the hoodie: it was about inspiring a whole new movement in manufacturing to improve quality and be based in the US. Our campaign goal was to inspire people about synthetic biology, so we made that a core feature of the video. Backers support a project and take a risk because the campaign is about more than just a product.

A good video is critical. Kickstarter says you can make it yourself, but we suck at videography, so we decided to outsource it, which turned out to be a great decision. (If you're looking for one, talk to Rick Symonds (*http://www.ricksymonds.com/contact*).) Doing this properly takes capital (in the range of thousands of dollars), but it also signals to backers that you are serious about the project if you have put your own money at risk before launch. Good music is important, and I spent two days reviewing clips on Audio Socket (*https://www.audiosocket.com/*) and The Music Bed (*http://www.themusicbed.com*). Shoot the video, get feedback on the draft, then reshoot and repeat.

Setting the target goal is hard—you want a low goal that looks attainable (and that you can get to the magic 20%), but not so low a goal that executing will be impossible. Momentum matters, and people like to back a winner, so the lower the goal you set, the better—most projects that reach their goal go significantly over it. We set our goal at the minimum amount necessary to still actually want to do the project. This approach means you can do an all-or-nothing campaign, as raising less than our goal would make it a struggle to execute.

Running the Campaign

We split the campaign into three phases:

1. Getting to 20% of our target funding goal by reaching our friends and family, who all knew the campaign was coming in advance and had promised to back within the first few days.

2. Getting to the goal of $65k by reaching out to the tech community/early adopters.

3. Going beyond that goal by reaching out to the mainstream/national press and to the broader public. (This phase was not well planned in advance, and what we did plan, for example, gardeners as our target segment, mostly failed.)

Within each group, we wanted to create a *surround sound* effect so that people in that segment saw the project three to four times in a week and hopefully decided to back it.

We launched at 9:30 a.m. Pacific time on a Tuesday; we selected this time because it gave us nearly the whole week to generate press and wasn't Monday morning, when people are just getting into their work week. The 20% target all knew the campaign was coming and what time it was expected to launch. We had emails ready to go to them and some friendly mailing lists (e.g., DIYbio and Bio-Curious) the second we went live. Around 20 people were anticipating the launch and ready to start sending messages to other people and share on social media.

We then posted to HackerNews, which didn't yield too many donations. However, it did result in people posting our campaign on other social news sites like reddit and Slashdot, resulting in some significant traffic over several days. Reddit was especially powerful—it brought more donations than the *New York Times* article!

Because I live in the Bay Area, I know someone at Techcrunch, which helped us get one of the journalists to cover the Glowing Plant story. We offered her an exclusive in exchange for Techcrunch holding off on publishing until after we launched the campaign, and that article led to many more (like *Gizmag, Fast Company, inhabitat (http://bit.ly/LVl5t3)*). We learned that all the leading blogs are read by other blogs who will want to write follow-up articles once you become a news story (this was also true with mainstream press, who all got in touch after the *New York Times* article). Thus the first story is the hardest to get, so network!

After that, it was mostly momentum and being prompt and responsive to incoming interest. Journalists are busy, and if they reached out, the least we could do was reply as soon as possible to schedule a time to talk. I worked 16 hours a day dealing with the onslaught, and we had two interns helping answer messages (we got hundreds per day). You can't run a campaign like ours part time or in addition to other activities.

One trick we learned was syndication. Obvious examples of this are tweets or Facebook shares of articles, but some news agencies also push articles out to other sites in syndication deals. For Singularity Hub, this led to five times the traffic from the original article, so it's always worth asking journalists to syndicate you.

Another thing we learned was the value of controversy, so don't be afraid of it. We had NGO's who didn't like our project pitch to mainstream news agencies for us, resulting in press like the *New York Times* piece, which we could never have gotten ourselves. You are not building consensus with your campaign—you are firing up your core supporters, your core believers, so much that they want to give you money even though you don't have a product yet. You want discussion and debate—that's what drives passion, and that's what drives social media. People share your project when they care, and controversy makes them care. Anything in synthetic bio will be controversial, so don't be afraid of that, but just remember to stick within clear ethical and lawful lines and develop a thick skin; on the Internet, haters are just going to hate, so accept it.

In the last days of the campaign, we used Hootsuite to schedule hourly tweets and Facebook messages counting down until the end. It's hard to track whether these worked, but they created urgency at the end to go along with the Kickstarter reminder email.

What Didn't Work

Lots of things went well with our campaign, but there were a few that didn't:

Stretch goals
> We hoped for success, but we didn't plan well enough for it with our stretch goals. I wish we had made them more frequent milestones and things that benefited all backers. Every project is different, but we should have outlined which additional experiments we could do with additional funds and linked those experiments to the stretch goals explaining how they would make the plant brighter. Still, I'm very excited about the glowing rose!

Duration
> Kickstarter recommends 30 days; we didn't listen and did 45 days. Most campaigns are a U-shaped (see our Kicktraq data (*http://bit.ly/1mtLcSA*)), with a huge peak at the start, then a long lull, then another rush at the end after Kickstarter sends the reminder email. If you extend the campaign, all you do is extend the lull, which means more stress until you get the funds.

Facebook ads

We tried running Facebook ads to drive traffic in the lull, but they brought in less money than we spent on them. However, I wish we could have targeted people who liked the page as well as their friends.

PR agency

We had a good conversion rate (2–8%, depending on source) and figured more traffic would lead to more conversions, so we hired a PR agency that contacted us. They got a few articles, but nothing compared to what we did ourselves. Journalists want to talk directly to the founders—you don't need a middleman.

Post-Campaign

We vaguely thought we'd plan the post-campaign during the actual campaign, which was a big mistake, as we didn't have time. The campaign is just crazy—we should have developed the strategy for that before launching. The most important thing is to plan where to direct traffic after the campaign ends because Kickstarter locks the page the second you finish. You have to set up those links ahead of time. If you will continue taking preorders, you want to decide that up front so that it can all be setup in advance. *http://trycelery.com* and *http://shopstarter.com* seem to be the leading platforms to help with that.

You also want to think about search engine optimization (SEO) post-campaign. We were encouraged to give a redirect link out to journalists (*http://www.glowing plant.com/kickstarter*) for the page rather than the actual page. That way, you can change it later and keep the SEO juice for yourself.

Another thing to plan for are credit card defaults (which is less of a problem on Indiegogo, which collects money upfront). These defaults come mostly from expired credit cards, and you can expect to spend the first week after the campaign chasing these people down to get new card details, so don't go on holiday too fast! After about one week, the campaign funding gets locked, and a week or so later, you get the funds. Our net proceeds were $432k, so about 89% of the total (the remainder is comprised of defaults, Amazon fees, and Kickstarter fees). Don't forget to budget this 11% cost in your planning!

Conclusion

The campaign was one of the most rewarding events in my career. It was an intense but incredible feeling to see the numbers go up and knowing people were putting their faith in us. We take the trust and responsibilities of the backers very seriously, and the plants are already glowing. We can't wait to ship them to everyone next summer!

We benefited from so many people's advice before launch, and we want to pay that forward to others. If you are planning a similar campaign, please do get in touch—we want to help! Just make sure to give us at least a couple weeks before your launch date to get back to you.

Appendix

For the geeks, here are our traffic stats:

- Video views: 358,000

- Backers: 8,433

- Funds pledged: $484,000

Referrer	Type	# Of Pledges	% Of Pledged	Pledged
Technology (Discover)	Kickstarter	1315	15.72	$76,075.14
Direct traffic (no referrer information)	External	1029	13.72	$66,386.69
Search	Kickstarter	1048	12.41	$60,044.07
singularityhub.com	External	498	6.3	$30,494.24
Popular (Discover)	Kickstarter	520	5.79	$28,045.50
Home popular	Kickstarter	435	4.73	$22,886
google.com	External	362	4.38	$21,216
glowingplant.com	External	186	4.1	$19,836.01
Staff Picks (Discover)	Kickstarter	316	3.53	$17,105
Home spotlight	Kickstarter	297	3.49	$16,901
Facebook	External	273	2.71	$13,126.50
Kickstarter user profiles	Kickstarter	144	1.84	$8,893
reddit.com	External	148	1.45	$7,026
nytimes.com	External	151	1.39	$6,717
boingboing.net	External	84	0.99	$4,777
techcrunch.com	External	92	0.96	$4,667
kurzweilai.net	External	90	0.95	$4,577
Starred	Kickstarter	91	0.93	$4,523
bbc.co.uk	External	93	0.86	$4,152
Home location	Kickstarter	74	0.82	$3,976
48-hour reminder email	Kickstarter	82	0.79	$3,846
Twitter	External	74	0.77	$3,747
popsci.com	External	69	0.67	$3,249
news.cnet.com	External	46	0.54	$2,598
Tag	Kickstarter	51	0.5	$2,442

Antony Evans is one of the founders of the Glowing Plant project. He led the first-ever crowdfunding campaign for a synthetic biology application, raising $484,000 from 8,433 backers on Kickstarter to create a glowing plant. He also co-runs The Container Lab, San Francisco's DIYbio lab. Evans received an MBA with distinction from INSEAD, an MA in maths from the University of Cambridge, and is a graduate of Singularity University's GSP program. He is both a Louis Frank and Oppidan scholar and worked for six years as a management consultant and project manager at Oliver Wyman and Bain & Company. Prior to this project, he cofounded the world's first pure mobile microfinance bank in the Philippines and launched the number one–ranked Android symptom checker mobile app in partnership with Harvard Medical School. Read more about the Kickstarter campaign (http://kck.st/1dfTdei).

Cheese, Art, and Synthetic Biology

An Interview with Christina Agapakis

Katherine Liu

Katherine Liu: **What can art and design teach us about biology and synthetic biology?**

Christina Agapakis: That's a great question. There are two different ways you can think about it: first as a way to reach different groups of people and have a different kind of conversation or debate around biotechnology. The second way that you could think about it is more interesting to me as a scientist, because I think using art and design helps us ask different questions and think about problems and technological solutions in different ways. To make a good technology, we need to be aware of both the biological and the cultural issues involved, and I think the intersection of art and design with science and technology helps us see those connections better.

KL: What kinds of projects have you done by combining art and biology?

CA: I'm really interested in bacteria and bacterial communities and how bacteria show us a different part of the world that we don't normally see. So a lot of the work I've been doing with art hasn't necessarily been about synthetic biology directly, but instead about how we interact with bacteria on our bodies and in our environment and how these relationships might change in the future as synthetic biology develops. So for example, although it came out of Synthetic Aesthetics (*http://synthe ticaesthetics.org*), which was about connecting synthetic biology with art and design, the cheese project (*http://agapakis.com/selfmade.html*) isn't really about the potential for genetic engineering to create synthetic biology technologies. Instead, we used cheese as a model for thinking about a much more basic form of biotechnology, how we can shape communities of bacteria to create these really fantastic and delicious products, and how our bodies and our food are these really fascinating ecosystems. Other projects I've been working on recently have been around more environmental issues. I've been isolating microbes from polluted water (*http://agapakis.com/encounters.html*) and from soils (*http://agapakis.com/dirt.html*) around California, using bacteria to understand how humans interact with the environment.

KL: **How did you first get involved with biotechnology?**

CA: I was really excited about biology in high school and then kind of obsessed with everything I learned about molecular biology and biochemistry in college. When I started working in a lab, I learned that a lot of basic biology experiments involve genetic engineering. To understand how genes work, people are moving genes around and understanding how things fit together, and I was excited to be learning those techniques and tools. But it wasn't until graduate school when I first met my advisor, Pam Silver, that I heard about synthetic biology. She's one of the leaders in the field, and she really inspired me to think about the things I had learned in my biology classes and in the lab not just as a way to learn more about how cells work, but also as a way to do engineering and to build useful things. That was really exciting for me, and Pam is great. So that's how I got into the field of synthetic biology!

KL: **What kind of trends do you see coming up in biotechnology?**

CA: I think the field is definitely maturing in some really interesting ways. In academia, we're seeing a lot more complexity in the kinds of projects people are working on. A lot of the projects that have been talked about for a while but have been a lot of really hard work to build are starting to come online, in particular projects like the Church lab's reprogrammed E. coli genome. I'm also really excited by what I see happening in terms of synthetic ecologies. You see more people working with communication between bacteria and engineering communities of bacteria to do things.

KL: **What do you think the future of synthetic biology is going to look like?**

CA: For me, I want to see it become more like biology—more messy, more like cheese making than like computer science. The analogies between computers and cells have been really interesting and have gotten people really excited about synthetic biology's potential, but I think what we're going to see is a transformation: a new paradigm as we learn how complicated things are inside a cell and where those analogies break down. We're going to be able to develop new ideas based around the ways that biology does things that are going to be more complex and more robust and able to adapt in interesting ways, and I think that's going to shift the way that we think about biotechnology.

KL: **I noticed that you've been an iGEM advisor. How can we bring biotechnology to younger students?**

CA: I think iGEM in particular has been really excited about getting college-level and now high school students to think about biology as an engineering platform. In high school, I was on the robotics team, and there were a lot of engineering competitions. But that hasn't really been there for biology yet, so iGEM is creating that same sort of idea of team-based projects around biology instead of around robots. It's been really great, and advising teams has been really fun for me and really rewarding. I was an advisor for the Harvard team for a couple of years, and I've been working with students at UCLA—this is the first year UCLA competed in iGEM. There were a couple of students who were really excited about starting a team and developing a project, and it was really fun and hard work to help them get this started. But it's been great to see students learn by doing and learn about what is possible with those tools by jumping into the lab.

KL: **What kinds of projects did the UCLA iGEM team work on?**

CA: This year, the UCLA team (*http://2013.igem.org/Team:UCLA*) was interested in phage, which are viruses that can kill bacteria, and they're interested in that relationship and the specificity between phage and bacteria. They found a really cool system where one phage uses recombination to generate a lot of diversity in the way that it interacts with the bacteria, so there's this natural system that the phage uses to accelerate evolution so it can interact with different things on the surface of the bacteria. They use that protein as a scaffold to do protein engineering, so they were looking at natural systems that created a lot of diversity and using those as a scaffold to generate diversity in vitro, in the lab to apply it to other cells.

KL: **I think an issue that a lot of students who go into biology face is whether to go into academia or industry right out of school. Why did you choose academia?**

CA: I wanted to learn forever—I was really excited about going to graduate school. I didn't know a lot about biotechnology and I'd never heard of synthetic biology; I was just excited about biology and chemistry and how things worked. I wanted to figure out how those things worked, so that's why I went into grad school right out of college. Now we see more and more that there's an overlap between the applied technology-building side of industry and the knowledge-building side of academia. There's a really interesting connection between making and knowing. In synthetic biology, that's really clear—as we make things, we understand more about them, and there's some really interesting crossovers happening between industry and academia, too.

KL: Do you have any advice for students who really want to study what you do, especially areas combining biology and art?

CA: The advice that I give to students is always just to follow what you're curious about, and read a lot. What I see a lot is students who are really curious and passionate about certain fields, but they actively stop themselves from learning more about it, from following this curiosity, because they think that it might not be useful or that it doesn't fit with the idea of what a good student or a good scientist would be interested in. I was an instructor for an art and science summer program at UCLA for high school students (*http://summer.artscicenter.com/*), and many of the students came in with an idea of what counted as science and what counted as art, and what they were good at or not good at. They would say, "I'm good at physics, and I only want to do physics," (or even worse, "I'm not good at math, I don't want to math and science"). But through the two weeks we were doing the program, they began to see the connections between what they were really excited about in physics or in art and other things, maybe in other fields of science or in fields of art and the humanities. So my advice is: don't be limited by what you think you're good at or what you're supposed to be good at, because some of the most interesting things you can learn come from the connections that you can make from looking a little bit outside of the path that you're on.

Influenced by her early fascination with the overlap between biotechnology, engineering, medicine, and design, Katherine Liu started a blog called Bioinnovate, where she publishes interviews that she conducts with leaders and entrepreneurs in the biotech space. Her interests include encouraging STEM education and expanding opportunities for young people. She can be reached at http://bioinnovate.tumblr.com.

Do Biohackers Dream of GM Sheep?

Ryan Bethencourt

Where do we go to dream? It's a question I asked myself as I heard the news about the FDA cracking down on 23andMe, a service that dared to dream outside of our regulatory system's confines. As we travel down new and unexplored paths, where do we go to move away from the understandably conservative thinking of our industry and think about biotechnologies that will, as Astro Teller from Google X labs says, make a 10x improvement to our world? I grew up reading mind-expanding sci-fi from Asimov, Sagan, Bear, Heinlein, Egan, Stross, and many others, and yet I couldn't find a list for those of us who wanted to dive deeper into the oddities and possible futures of the technology of life. So after a discussion with Kyle Taylor, the head scientist for Glowing Plants, in which we discussed one future possibility of engineering crops to grow on the surface of oceans to feed the hungry billions of the world, I thought it would be worth reaching out across the Web to compile a list of other future possible biotechnologies.

Hopefully some of the books and short stories in this list will help inspire the next generation of biohackers, scientists, and DIY biologists.

Short Stories

- *Ribofunk*, by Paul Di Filippo, is an excellent short-story collection.
- *BioPunk*, edited by Ra Page, is a short-story anthology penned by different authors. The short stories were commissioned to be based on actual current research, and authors were paired up with a scientist or ethicist that fact-checked the story and also provided a commentary. The stories were of varying quality and relevance, but it's definitely an interesting concept.

- *Roo'd* (*http://www.josh.is/rood-a-cyberpunk-novel/*), by Joshua Klein. Cyberpunk meets DIYbio. It's a self-published, CC book, and it's better than some of the best authors in the same genre.
- *Chaff*, by Greg Egan, is a biochemical warfare story that takes place in the Amazon.
- For something a little different, "Genocide Man" is a webcomic story based on open source biotechnology that enabled various ideological groups to create designer plagues and super soldiers, wiping out most of the world's population. The titular character is a rogue law enforcement agent whose job is to stop carriers of dangerous ideologies from committing genocide again by killing them all first. He has a portable lab the size of a large suitcase that can produce viruses tailored to only kill specific people or gene lines. Sounds intense and post apocalyptic, but could be a fun read!

Long Reads

- *Frankenstein*, by Mary Shelley: the original biohacker sci-fi anthem, and it's definitely worth a read!
- *Nexus and Crux*, by Ramez Naam: hard bionanotech, and totally recommended.
- *The Windup Girl*, by Paolo Bacigalupi. He does for biotech what William Gibson did for cyberspace with Neuromancer in 1984—it has changed the way some of us think about biology as a political technology completely. And it's a great story (it won the Hugo and Nebula awards).
- *Oryx and Crake*, by Margaret Atwood, is a dystopian novel set in a not-so-distant apocalyptic future in which mankind has been eradicated by the jealousy of one man.
- *Embassytown*, by China Miéville, has a strong organic tech strand and was highly recommended.
- *Rule 34*, by Charles Stross, has a synth-bio subplot.
- *Do Androids Dream of Electric Sheep?* by Philip K. Dick is the book that inspired *Blade Runner*. It's a must-read classic!
- *Darwin's Radio*, by Greg Bear, is a fascinating idea based on the development of an endogenous virus that controls humanity's evolution while in the womb.

- *Blood Music*, by Greg Bear. This is what happens when you add human-level intelligence (or beyond) to cells. The world gets really weird and intriguing!

- *Altered Carbon*, by Richard K. Morgan. A true inheritor of the cyberpunk genre and pure, hard-edged biotech sci-fi. If you want to glimpse into the possible future laced with substrate independent minds, genetically tailored and built bodies, and a world in which body swapping is the norm, this is *the* book!

- *Accelerando*, by Charles Stross. Lots and lots of genetic engineering as we accelerate toward and past the technological singularity. It's a wild ride!

- *Methuselah's Children* and *Time Enough for Love*, by Robert Heinlein. These two stories capture the earlier mentions of applying the tools of natural selection and Mendelian genetics to extend human lifespan into the hundreds of years and beyond.

- *Hyperion Cantos*, by Dan Simmons, is a fascinating trip through the possible future of humanity's directed evolution leading to a rainbow of different human races, many of which are designed to be able to survive in deep space.

- *Blindsight*, by Peter Watts. Nominated for a Hugo award, a Campbell award, and a Locus Science Fiction award, this book covers both extraterrestrial biology and an offshoot of humanity.

- *The Breeds of Man*, by F. M. Busby. A genetically engineered cure to the ravages of AIDS that leads to sterility, so a new genetically engineered version of humanity is developed, with a twist!

- *The Stars My Destination*, by Alfred Bester, had some very interesting biotech details scattered through it and a sci-fi adaptation of the *Count of Monte Cristo*.

- *Ware Tetrology*, by Rudy Rucker, is huge for biotech: it melds AI, evolution, tech, and biotech.

- *Deathworld*, by Harry Harrison, had some mind-bending evolutionary implications in it.

- *Starseed Transmission*, by Ken Carey, paints an amazing picture of a distant future (or parallel reality) where biotech exists on a much different level.

- *The Uplift War*, by David Brin, was captivating, with the main concept focused on galactic civilizations using biotechnology to find near-sapient creatures and force them into full sentience.

- *The White Plague*, by Frank Herbert, was one of the first books that featured a tailored viral plague, which inspired other authors to expand on his visions of the future.

- *The Child Garden*, by Geoff Ryman Page, explores the positive use of viruses, to the point that biotech becomes the only tech.

- *Diaspora*, by Greg Egan, is a journey through post-humanity's future, when humanity has speciated into three distinct groups. It focuses on the nature of life and intelligence in a post-human future and is a must-read for those who want to glimpse what might be for minds that may exist in a substrate, independent way.

- *Rainbows End*, by Vernor Vinge, has some very cool visions of large-scale genetic and molecular engineering.

Where do you go to dream of biotech futures to come? Tweet me (*@ryanbethencourt* (*https://twitter.com/ryanbethencourt*)) if you have any additional suggestions, and I hope you enjoy a few of the short stories and books on this list!

A special thanks to everyone who contributed suggestions and commentary, including: Andrew Hessel, Nathan McCorkle, Alexander Hollins, Gunther Mulder, Matthew Pocock, Andreas Stuermer, Will Canine, Paul Schroeer-Hannemann, Philipp Boeing, Mackenzie Cowell, Dr. Brian, Pat Moss, Jen Zariat, Jake Raden, Alex Hoekstra, Chris Folk, Xander Honkala, Kent Kemmish, Colin Ho, Jonathan Yankovich, Antony Evans, Ty Larson, Damon Millar, Lianne and Stephen Holmes, Richard Hodkinson, Gabriel Stempinski, Justin Dormandy, Keith Causey, and Alan Schunemann.

@ryanbethencourt (https://twitter.com/ryanbethencourt) is the CEO of Berkeley Biolabs and a managing partner of LITMUS Clinical Research. He's currently working to accelerate innovations in biotechnology and medicine through biohacking, open innovation, and collaboration. His primary areas of expertise are human translational medicine, genetics, longevity research, and business development.

Community Announcements

Toronto DIYbio

Toronto DIYbio has gained tremendous momentum over the past year and is preparing to introduce itself as Canada's newest biohackerspace. Currently hosting monthly organizational meetings, lectures on synthetic biology, and a Molecular Biology of the Cell study group, they are welcoming all guests or contributors to come and be a part of this exciting process. With support from local hackerspace HackLab.TO and art-science gallery Action Potential (*http://www.actionpotential lab.ca*), they are quickly outgrowing their shared accommodations and will be expanding to an independent laboratory space in 2014. Meeting information can be found at Meetup.com (*http://www.meetup.com/DIYbio-Toronto/*), and stay tuned to the DIYbio Toronto home page (*http://diybio.to/*) and the DIYbio Twitter feed (*https://twitter.com/biybioto*) for all information about recent developments and upcoming fundraising campaigns.

SynBioaxlr8r

SynBioaxlr8r (*http://www.synbioaxlr8r.com*) has been setup to overcome some of the hurdles that entrepreneurs encounter on their way to success in the exciting field of synthetic biology. This 90-day program is committed to idea-stage startups, offering them lab space and $30,000 in funding as well as world-class mentoring, advice, and support from the top experts in the industry. The synbio entrepreneurs have access to top outsourced DNA synthesis and an open source computer programming language that compiles into DNA for rapid prototyping. The best and brightest mentors from all over the world help curate and cultivate every aspect of each business. This fuels groundbreaking ideas in the areas of medicine, energy, agriculture, and manufacturing.

Over the course of the program, teams gain enormous validation and can transform themselves into functional companies. They have an opportunity to immerse themselves in an intensive startup community of like-minded entrepreneurs. The venue is based in University College, Cork, and the program will run over the summer of 2014. Ten outstanding teams from around the world will be selected to participate in the program. This exciting accelerator has already attracted huge interest from leaders in the field worldwide. One of the main objectives of the program is to have Cork become a world-leading center for synthetic biology. This is just the beginning of something huge. This is a revolutionary area where the possibilities are endless. Synbio is fast paced, exciting, and a perfect fit for an accelerator program, and the program is a platform that has the potential to change the world for the better.

Ask a Biosafety Expert

DIYbio.org and the Synthetic Biology Project at the Woodrow Wilson Center launched the Ask a Biosafety Expert service (*http://ask.diybio.org/*) in January 2013 to provide the emerging DIYbio community with free and timely access to professional biosafety advice. The aim of the Ask a Biosafety Expert service is to provide advice from experts about laboratory safety, disposal, and other bio-related issues to members of the DIYbio community. Questions submitted to *http://diybio.org* are sent to a panel of professional biosafety experts for prompt and user-friendly feedback. The responses are then posted and archived on the public website, providing a growing resource for DIYbio community.

Berkeley BioLabs

 Berkeley BioLabs is focused on transforming the way therapeutics, medical devices, diagnostics, and green and industrial biotech are developed, through a collaborative shared lab for scientists and entrepreneurs. The primary goal is to focus on lowering the overall development cost and support idea to proof-of-concept–stage projects through low-cost access to equipment, mentoring, and in some cases financial support.

We are launched and ready for awesome scientists to populate the space. Check us out at *http://berkeleybiolabs.com/* and please contact us for more information at *contact@berkeleybiolabs.com*.

DIYbio and iGEM

 iGEM just opened the competition to community lab teams in 2014! In order for DIY teams to participate, they must obtain the parts from the biobricks standardized parts registry, pay the registration fees, have two PI leaders, and find a company to host them. The DIY teams will have the opportunity to compete against all other teams for the grand prize. Get excited, start brainstorming, recruit awesome team members, secure funding, and create some amazing science.

For more information, go to *http://diybio.org/2013/11/06/diy-igem/* and send questions to *diy@igem.org*.

Lightning Source UK Ltd.
Milton Keynes UK
UKOW06f2110120116

266255UK00001B/1/P.